TRANSITIONS TO ALTERNATIVE TRANSPORTATION TECHNOLOGIES—PLUG-IN HYBRID ELECTRIC VEHICLES

Committee on Assessment of Resource Needs for Fuel Cell and Hydrogen Technologies

Board on Energy and Environmental Systems
Division on Engineering and Physical Sciences

NATIONAL RESEARCH COUNCIL
OF THE NATIONAL ACADEMIES

THE NATIONAL ACADEMIES PRESS
Washington, D.C.
www.nap.edu

THE NATIONAL ACADEMIES PRESS • 500 Fifth Street, N.W. • Washington, DC 20001

NOTICE: The project that is the subject of this report was approved by the Governing Board of the National Research Council, whose members are drawn from the councils of the National Academy of Sciences, the National Academy of Engineering, and the Institute of Medicine. The members of the committee responsible for the report were chosen for their special competences and with regard for appropriate balance.

This study was supported by Contract DE-AT01-06EE11206, TO#18, Subtask 3, between the National Academy of Sciences and the U.S. Department of Energy. Any opinions, findings, conclusions, or recommendations expressed in this publication are those of the author(s) and do not necessarily reflect the views of the organizations or agencies that provided support for the project.

International Standard Book Number 13: 978-0-309-14850-4
International Standard Book Number 10: 0-309-14850-2
Library of Congress Control Number: 2010925717

Cover: Images (adapted) courtesy of California Cars Initiative (left) and U.S. Department of Energy (right).

Available in limited supply from:

Board on Energy and Environmental Systems
National Research Council
500 Fifth Street, N.W.
Keck W934
Washington, DC 20001
(202) 334-3344

Additional copies for sale from:

The National Academies Press
500 Fifth Street, N.W.
Lockbox 285
Washington, DC 20055
(800) 624-6242 or (202) 334-3313
(in the Washington metropolitan area)
Internet: http://www.nap.edu

Printed in the United States of America

THE NATIONAL ACADEMIES

Advisers to the Nation on Science, Engineering, and Medicine

The **National Academy of Sciences** is a private, nonprofit, self-perpetuating society of distinguished scholars engaged in scientific and engineering research, dedicated to the furtherance of science and technology and to their use for the general welfare. Upon the authority of the charter granted to it by the Congress in 1863, the Academy has a mandate that requires it to advise the federal government on scientific and technical matters. Dr. Ralph J. Cicerone is president of the National Academy of Sciences.

The **National Academy of Engineering** was established in 1964, under the charter of the National Academy of Sciences, as a parallel organization of outstanding engineers. It is autonomous in its administration and in the selection of its members, sharing with the National Academy of Sciences the responsibility for advising the federal government. The National Academy of Engineering also sponsors engineering programs aimed at meeting national needs, encourages education and research, and recognizes the superior achievements of engineers. Dr. Charles M. Vest is president of the National Academy of Engineering.

The **Institute of Medicine** was established in 1970 by the National Academy of Sciences to secure the services of eminent members of appropriate professions in the examination of policy matters pertaining to the health of the public. The Institute acts under the responsibility given to the National Academy of Sciences by its congressional charter to be an adviser to the federal government and, upon its own initiative, to identify issues of medical care, research, and education. Dr. Harvey V. Fineberg is president of the Institute of Medicine.

The **National Research Council** was organized by the National Academy of Sciences in 1916 to associate the broad community of science and technology with the Academy's purposes of furthering knowledge and advising the federal government. Functioning in accordance with general policies determined by the Academy, the Council has become the principal operating agency of both the National Academy of Sciences and the National Academy of Engineering in providing services to the government, the public, and the scientific and engineering communities. The Council is administered jointly by both Academies and the Institute of Medicine. Dr. Ralph J. Cicerone and Dr. Charles M. Vest are chair and vice chair, respectively, of the National Research Council.

www.national-academies.org

COMMITTEE ON ASSESSMENT OF RESOURCE NEEDS FOR FUEL CELL AND HYDROGEN TECHNOLOGIES

MICHAEL P. RAMAGE, NAE,[1] *Chair,* ExxonMobil Research and Engineering Company (retired), Moorestown, New Jersey
RAKESH AGRAWAL, NAE, Purdue University, West Lafayette, Indiana
DAVID L. BODDE, Clemson University, Clemson, South Carolina
DAVID FRIEDMAN, Union of Concerned Scientists, Washington, D.C.
SUSAN FUHS, Conundrum Consulting, Hermosa Beach, California
JUDI GREENWALD, Pew Center on Global Climate Change, Washington, D.C.
ROBERT L. HIRSCH, Management Information Services, Inc., Alexandria, Virginia
JAMES R. KATZER, NAE, Massachusetts Institute of Technology, Washington, D.C.
GENE NEMANICH, ChevronTexaco Technology Ventures (retired), Scottsdale, Arizona
JOAN OGDEN, University of California, Davis, Davis, California
LAWRENCE T. PAPAY, NAE, Science Applications International Corporation (retired), La Jolla, California
IAN W.H. PARRY, Resources for the Future, Washington, D.C.
WILLIAM F. POWERS, NAE, Ford Motor Company (retired), Boca Raton, Florida
EDWARD S. RUBIN, Carnegie Mellon University, Pittsburgh, Pennsylvania
ROBERT W. SHAW, JR. Aretê Corporation, Center Harbor, New Hampshire
ARNOLD F. STANCELL,[2] NAE, Georgia Institute of Technology, Greenwich, Connecticut
TONY WU, Southern Company, Wilsonville, Alabama

Consultant

JAMES CANADA

Project Staff

Board on Energy and Environmental Systems

ALAN CRANE, Study Director
JAMES ZUCCHETTO, Director, BEES
JONATHAN YANGER, Senior Project Assistant

NAE Program Office

PENELOPE GIBBS, Senior Program Associate

[1]NAE, National Academy of Engineering.
[2]Resigned from the committee June 2009.

BOARD ON ENERGY AND ENVIRONMENTAL SYSTEMS

[1]NAE, National Academy of Engineering.
[2]Term ended September 30, 2009.
[3]NAS, National Academy of Sciences.

Preface

The Committee on Assessment of Resource Needs for Fuel Cell and Hydrogen Technologies completed its report *Transitions to Alternative Transportation Technologies—A Focus on Hydrogen* (The National Academies Press, Washington, D.C.) in 2008. Subsequently, the U.S. Department of Energy requested the National Research Council (NRC) to expand that analysis to plug-in hybrid electric vehicles (PHEVs). The committee reconvened to examine the issues associated with PHEVs and wrote this report in response to that additional task.

The nation has only a few options for making great reductions in its dependence on oil and emissions of carbon dioxide, the main greenhouse gas, from the transportation sector. Hydrogen fuel cell vehicles are one, and electric vehicles are another. Both have great potential but also serious disadvantages and uncertainties. In particular, costs for both are currently very high, and both have limited range.

In comparison, PHEVs have some attractive characteristics. Unlike hydrogen fuel cell vehicles, they can be deployed in the marketplace without simultaneously building an infrastructure to supply the energy to operate them, and unlike all-electric battery vehicles, drivers will not have to worry about charging the batteries on a long trip. However, PHEVs have their own limitations, as discussed in this report.

It is unusual for the NRC to reconvene a committee organized for one purpose to investigate another, but this is an unusual committee in another way, too. I have never worked with a committee that was so dedicated, knowledgeable, and talented. This entire additional task has taken about 6 months, an extraordinarily fast pace for a complex issue. The committee members have my deepest appreciation. The project also was very fortunate in having as its study director Alan Crane, who contributed immeasurably with his experience and expertise and his ability to keep the whole process moving on schedule.

The committee operated under the auspices of the NRC Board on Energy and Environmental Systems and is grateful for the able assistance of James Zucchetto and Jonathan Yanger of the NRC staff, and Penelope Gibbs of the National Academy of Engineering Program Office staff.

Michael P. Ramage

Acknowledgments

The Committee on Assessment of Resource Needs for Fuel Cell and Hydrogen Technologies is grateful to the many individuals who contributed their time and efforts to this National Research Council (NRC) study. The presentations at committee meetings provided valuable information and insights that enhanced the committee's understanding of the technologies and barriers involved. The committee thanks the following individuals and companies for their briefings and information:

Shinichi Abe, Toyota Motor Corporation,
Dick Cromie, Southern California Edison,
Bob Graham, Southern California Edison,
Dave Howell, U.S. Department of Energy,
Tien Nguyen, U.S. Department of Energy,
Phil Patterson, U.S. Department of Energy,
Bill Reinert, Toyota Motor Sales, USA, Inc.,
Sandy Thomas, H2Gen,
Mark Verbrugge, General Motors,
David Vieau, A123 Systems,
Michael Wang, Argonne National Laboratory,
Jake Ward, U.S. Department of Energy,
Compact Power, Inc.
Delphi Corporation,
DENSO International America, Inc., and
Ford Motor Corporation.

This report has been reviewed in draft form by individuals chosen for their diverse perspectives and technical expertise, in accordance with procedures approved by the NRC's Report Review Committee. The purpose of this independent review is to provide candid and critical comments that will assist the institution in making its published report as sound as possible and to ensure that the report meets institutional standards for objectivity, evidence, and responsiveness to the study charge. The review comments and draft manuscript remain confidential to protect the integrity of the deliberative process. We wish to thank the following individuals for their review of this report:

Paul Blumberg, Consultant,
Andrew Brown, Delphi Corporation,
Doug Chapin, MPR Associates,
John German, International Council for Clean Transportation,
Charles Goodman, Consultant,
Paul Gray, Massachusetts Institute of Technology,
Daniel Greenbaum, Health Effects Institute,
Trevor Jones, ElectroSonics Medical, Incorporated,
Maryann Keller, Maryann Keller and Associates, and
Brijesh Vyas, LGS Innovations, Limited Licensing Corporation.

Although the reviewers listed above have provided many constructive comments and suggestions, they were not asked to endorse the conclusions or recommendations, nor did they see the final draft of the report before its release. The review of this report was overseen by Elisabeth M. Drake (NAE), Massachusetts Institute of Technology. Appointed by the National Research Council, she was responsible for making certain that an independent examination of this report was carried out in accordance with institutional procedures and that all review comments were carefully considered. Responsibility for the final content of this report rests entirely with the authoring committee and the institution.

Contents

[1]Appendix F was added to this report after release of the prepublication version to clarify how the cost estimates were made.

Tables, Figures, and Boxes

TABLES

FIGURES

BOXES

Summary

The nation has compelling reasons to reduce its consumption of oil and emissions of carbon dioxide. Plug-in hybrid electric vehicles (PHEVs) promise to contribute to both goals by allowing some miles to be driven on electricity drawn from the grid, with an internal combustion engine that kicks in when the batteries are discharged. However, while battery technology has made great strides in recent years, batteries are still very expensive.

This report builds on a 2008 National Research Council (NRC) report on hydrogen fuel cell vehicles (HFCVs).[1] In accordance with the committee's statement of task, the present report

1. Reviews the current and projected technology status of PHEVs.
2. Considers the factors that will affect how rapidly PHEVs could enter the marketplace, including the interface with the electric transmission and distribution (T&D) system.
3. Determines a maximum practical penetration rate for PHEVs consistent with the time frame and factors considered in the 2008 Hydrogen Report.
4. Incorporates PHEVs into the models used in the hydrogen study to estimate the costs and impacts on petroleum consumption and carbon dioxide (CO_2) emissions.

TECHNOLOGY STATUS

Vehicle Technologies and Batteries

A variety of PHEV configurations and electric driving ranges are under consideration by vehicle manufacturers. This report considers two vehicles. One, the PHEV-10, uses hybrid electric vehicle (HEV) technology similar to that used in the Toyota Prius. However, it has a larger battery than an HEV to allow 10 miles of driving powered by electricity only and a gasoline engine that drives the wheels in parallel with the electric motor when power demand is high or the batteries are discharged. The other vehicle, the PHEV-40, is similar to the Chevrolet Volt. It has a 40-mile electric range, a larger electric motor, and a much larger battery than the PHEV-10. In the PHEV-40, the electric motor provides all the propulsion; the gasoline engine drives a generator that powers the motor and keeps the batteries charged above some minimum level.

Batteries are the key determinant of the cost and electric driving range of PHEVs. All proposed PHEVs will use lithium-ion (Li-ion) batteries, similar to the technology now used in laptop computers, power tools, and other small devices. Several Li-ion chemistries are under development with the objective of optimizing performance for automotive propulsion. None yet meet all essential goals for cost, battery life, and weight. Cost is expected to be the most difficult goal.

The cost to the manufacturer of producing the first generation of the PHEV-10 (2010-2012) is expected to be about \$5,500 to \$6,300 more than that of the equivalent conventional midsize car (nonhybrid), including \$2,500 to \$3,300 for the battery pack. Similarly, the PHEV-40 with a \$10,000 to \$14,000 battery pack would cost about \$14,000 to \$18,000 more. These cost differences would be smaller if the PHEVs were compared to equivalent HEVs, but the fuel savings also would be smaller.

Costs will decline with technology improvements and economies of scale, but Li-ion batteries based on similar technology are already being produced in great numbers and are well along their learning curves. The steep early drop in cost often experienced with new technologies is not likely. The incremental cost to manufacture these vehicles is expected to decline by about one third by 2020 but only slowly thereafter, as listed in Table S.1.

[1]National Research Council, *Transitions to Alternative Transportation Technologies—A Focus on Hydrogen*, Washington, D.C.: The National Academies Press, 2008, hereinafter referred to as the 2008 Hydrogen Report.

TABLE S.1 Estimated Future PHEV Incremental Costs

	2011	2015	2020	2030
PHEV-40	14,100-18,100	11,200-14,200	9,600-12,200	8,800-11,000
PHEV-10	5,500-6,300	4,600-5,200	4,100-4,500	3,700-4,100

NOTE: These are the incremental costs to manufacture the vehicle itself, relative to a conventional (nonhybrid) vehicle. They do not include engineering, overhead, or other costs, or profit, and thus are not the total incremental prices to the customer. Costs for 2011 are based on low battery production rates in response to contracts initiated about 2 years earlier. Ranges represent probable and optimistic assessments of battery technology progress. Additional detail on the committee's analysis of battery-pack cost can be found in Appendix F, which was added to this report after release of the prepublication version to clarify how the estimates were made.

It is possible that breakthroughs in battery technology will greatly lower the cost. At this point, however, it is not clear what sorts of breakthroughs might become commercially viable. Furthermore, even if they occur within the next decade, they are unlikely to have much impact before 2030, because it takes many years to get large numbers of vehicles incorporating new technology on the road.

Electric Power Infrastructure Issues

PHEVs replace gasoline with electricity for some of the miles driven. The electricity will first have to be generated and then delivered to a PHEV through the electric grid. This raises two issues: (1) whether sufficient generation, transmission, and distribution capacity will be available to serve this additional load and (2) how the emissions from the additional electricity generation compare with the emissions from the gasoline not consumed.

Grid capacity will be available to charge millions of PHEVs if they are charged at night. Power demand varies during the day, peaking during the afternoon and reaching a low point after midnight. It also varies over the year, with demand highest on summer afternoons because of air conditioning loads. Parts of the U.S. electric power system are at full capacity during these hours of highest demand, and additional loads could threaten reliability unless new capacity is added. At night, however, the system may operate at less than 50 percent of capacity, and the cost of producing electricity is much lower than during peak hours. Drivers paying a constant rate per kilowatt-hour of electricity are likely to charge their vehicles whenever they have convenient access to an electric outlet, potentially increasing electricity demand during peak hours. Smart meters with time-of-use pricing would be one way of encouraging drivers to delay charging until electricity demand is lower.

Generating electricity to replace the gasoline that a car would have used emits some greenhouse gases (GHG), especially CO_2. About half the nation's electricity is produced from coal-fired power plants, which are large emitters of CO_2. However, the overall efficiency of electric vehicles is greater than that of conventional vehicles, so emissions may be reduced to some extent. Large savings on emissions will require decarbonizing the electric system, such as by using nuclear power or renewable energy generation or by capturing and sequestering the CO_2 emitted by fossil fuel plants.

SCENARIOS

Penetration rates for the PHEV-10 and the PHEV-40 were compared to a Reference Case that assumes high oil prices and fuel economy standards specified by the Energy Independence and Security Act of 2007 (with modest increases after 2020, when those standards level off), as described in the 2008 Hydrogen Report. The Maximum Practical scenario is the fastest rate at which the committee concluded that PHEVs could penetrate the market considering various manufacturing and market barriers; it leads to about 40 million PHEVs by 2030 in a fleet of about 300 million vehicles.[2] A more probable scenario leads to about 13 million PHEVs by 2030. Figure S.1 shows the number of PHEVs on the road at the two rates.

Figure S.2 shows the impact on gasoline use relative to the Reference Case when each of the two PHEV types is introduced at the Maximum Practical rate into a high-efficiency fleet. The Efficiency Case fleet, based on Case 2 from the 2008 Hydrogen Report, includes conventional nonhybrid vehicles and HEVs only. All cases give results similar to the Reference Case until after 2020, because it takes many years for a sufficient number of new vehicles to penetrate the market to have an impact. By 2030, the Efficiency and PHEV cases show gasoline consumption well below the Reference Case. PHEV-10 closely follows the Efficiency Case until 2040, after which it shows some additional benefit. PHEV-40 shows benefits relative to the Efficiency Case after 2025.

Figure S.3 shows the well-to-wheels GHG emissions of the light-duty vehicle fleet for the PHEV scenarios and compares them to the Reference Case. PHEVs show less improvement in GHG emissions than in gasoline consumption because of the additional emissions from electricity generation. If carbon emissions from the electric sector are limited, the reductions in Figure S.3 would be greater, almost following the reductions in gasoline use in Figure S.2.

[2]This scenario is based on the Hydrogen Success scenario in the 2008 Hydrogen Report but moved up 3 years because battery technology is more nearly ready for commercialization than fuel cells.

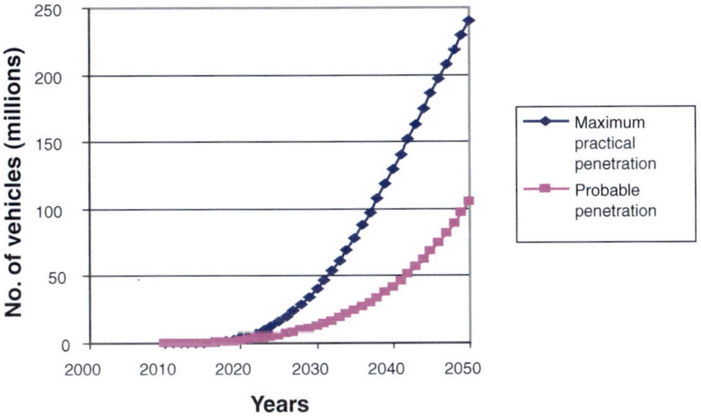

FIGURE S.1 Projections of number of PHEVs in the U.S. light-duty fleet.

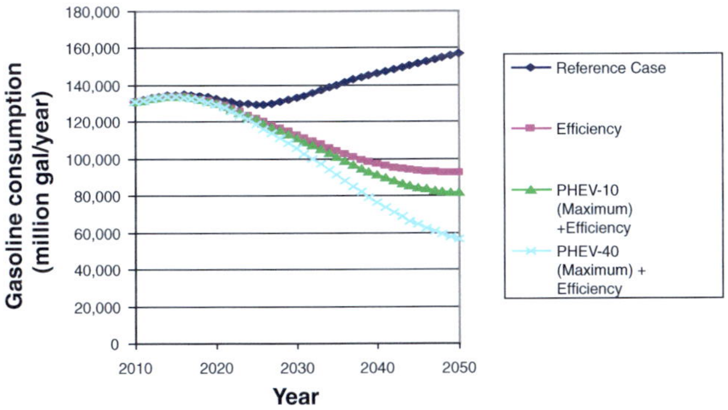

FIGURE S.2 Gasoline use for PHEV-10s and PHEV-40s introduced at the Maximum Practical rate and the Efficiency Case from the 2008 Hydrogen Report.

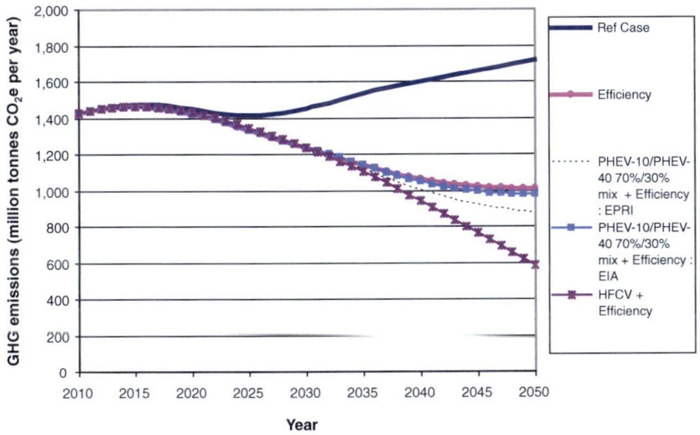

FIGURE S.3 GHG emissions for cases combining high-efficiency conventional vehicles and HEVs with mixed PHEV or HFCV vehicles for the two different grid mixes.

The PHEV projection cases considered only the impact of a given number of PHEVs regardless of cost. PHEVs will be expensive relative to conventional vehicles, largely because the batteries are costly. They are cheaper to operate (driving costs per mile are less than for conventional vehicles), and eventually vehicle costs may decline sufficiently to achieve life-cycle cost competitiveness, as shown in Tables S.1 and S.2. A transition period with substantial policy intervention and/or financial assistance for buyers from government and possibly manufacturers will be necessary to support either of the penetration scenarios in Figure S.1 until the higher costs of PHEVs are balanced by their fuel savings. The break-even year is defined here as the year when the fuel savings of the entire fleet of PHEVs equals the subsidies required that year to make PHEVs appear cost-competitive to potential buyers relative to conventional vehicles.

Transition costs will depend on how fast vehicle costs decline and how fast PHEVs penetrate the market. Table S.2 shows the break-even year and transition cost for the PHEV-40 for three Maximum Practical penetration scenarios: for the committee's optimistic assessment of technical progress; if DOE's goals for costs are met by 2020; and if oil prices are much higher than assumed for the base case. PHEV-40s achieve breakeven in 2040 for the committee's Optimistic technical progress, but in 2024 if DOE's goals are achieved, illustrating the potential importance of technology breakthroughs. Similarly, the required subsidies are much lower if oil prices are very high. PHEV-10s achieve breakeven much sooner and with much lower subsidies when analyzed on a basis comparable to PHEV-40s, but also provide lower oil and carbon emission benefits. The final two columns of Table S.2 show results for a mix of PHEV-40s and PHEV-10s, which are between those of each type analyzed alone, and for a slower growth rate with less optimistic technological progress.

Finally, the committee included combinations of technologies to reduce oil consumption in the light-duty vehicle fleet, as was done in the 2008 Hydrogen Report. Advanced conventional vehicles (including HEVs) operating in part on biofuels could cut oil consumption by more than 60 percent by 2050, as shown in Figure S.4. Replacing some of those HEVs with PHEVs, especially PHEV-40s, could reduce consumption to even lower levels. Employing HFCVs instead of PHEVs, however, could eliminate oil use in the light-duty vehicle fleet.

RESULTS AND CONCLUSIONS

1. Lithium-ion battery technology has been developing rapidly, especially at the cell level, but costs are still high, and the potential for dramatic reductions appears limited. Assembled battery packs currently cost about $1,250 to $1,700 per kWh of usable energy ($625 to $850/kWh of nameplate energy). A PHEV-10 will require about 2.0 kWh and a PHEV-40 about 8 kWh even after the batteries have

TABLE S.2 PHEV Transition Times and Costs

	PHEV-40	PHEV-40	PHEV-40 High Oil[a]	PHEV-10	30/70% PHEV-40/10 Mix	
Penetration Rate:	Maximum Practical	Maximum Practical	Maximum Practical	Maximum Practical	Maximum Practical	Probable
Technical Progress:	Optimistic	DOE Goal[b]	Optimistic	Optimistic	Optimistic	Probable
Break-even year[c] (annual cash flow = 0)	2040	2024	2025	2028	2032	2034
Cumulative subsidy to break-even year (billion $)[d]	408	24	41	33	94	47
Cumulative vehicle retail price difference until the break-even year (billion $)[e]	1,639	82	174	51	363	—
Number of PHEVs sold to break-even year (millions)	132	10	13	24	48	20

[a]Assumes oil costs twice that in the base case, or $160/bbl in 2020, giving results similar to meeting DOE's cost goals.

[b]Assumes DOE technology cost goal ($300/kWh) for the PHEV-40 is met by 2020, showing the importance of technology breakthroughs as discussed in Chapter 2 and Appendix F. Reducing costs this rapidly would significantly reduce subsidies and advance the break-even year relative to the Optimistic Technical Progress cases.

[c]Year when annual buydown subsidies equal fuel cost savings for fleet.

[d]Does not include infrastructure costs for home rewiring, distribution system upgrades, and public charging stations which might average over $1000 per vehicle.

[e]Cost of PHEVs minus the cost of Reference Case cars.

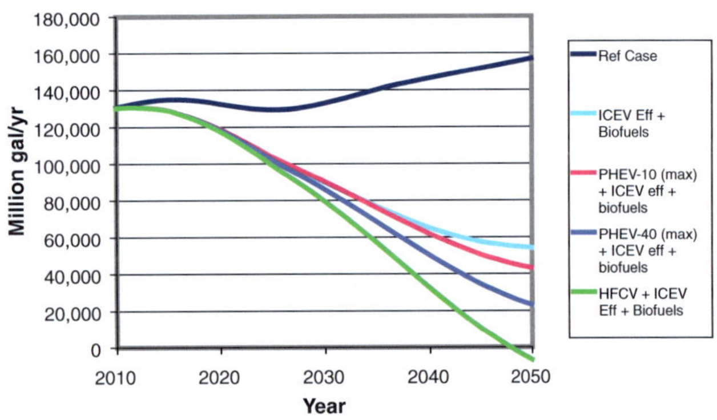

FIGURE S.4 Gasoline consumption for scenarios that combine conventional vehicle efficiency, PHEVs, biofuels, and HFCVs.

undergone expected degradation over time. Costs are expected to decline by about 35 percent by 2020 but more slowly thereafter. Projections of future battery pack costs are uncertain, as they depend on the rate of improvements in battery technology and manufacturing techniques, potential breakthroughs in new technology, possible relaxation of battery protection parameters as experience is gained, and the level of production, among other factors. Further research is needed to reduce costs and achieve breakthroughs in battery technology.

2. Costs to a vehicle manufacturer for a PHEV-40 built in 2010 are likely to be about $14,000 to $18,000 more than an equivalent conventional vehicle, including a $10,000 to $14,000 battery pack. The incremental cost of a PHEV-10 would be about $5,500 to $6,300, including a $2,500 to $3,300 battery pack. In addition, some homes will require electrical system upgrades, which might cost more than $1,000. In comparison, the incremental cost of an HEV might be $3,000.

3. PHEV-40s are unlikely to achieve cost-effectiveness before 2040 at gasoline prices below $4.00 per gallon, but PHEV-10s may get there before 2030. PHEVs will recoup some of their incremental cost, because a mile driven on electricity will be cheaper than a mile on gasoline, but it is likely to be several decades before lifetime fuel savings start to balance the higher first cost of the vehicles. Subsidies of tens to hundreds of billions of dollars will be needed for the transition to cost-effectiveness. Higher oil prices or rapid reductions in battery costs could reduce the time and subsidies required to attain cost-effectiveness.

4. At the Maximum Practical rate, as many as 40 million PHEVs could be on the road by 2030, but various factors (e.g., high costs of batteries, modest gasoline savings, limited availability of places to plug in, competition from other vehicles, and consumer resistance to plugging in virtually every day) are likely to keep the number lower. The Maximum Practical rate depends on rapid technological progress, increased government support, and consumer acceptance. A more realistic penetration rate would result in 13 million PHEVs by 2030 out of about 300 million vehicles on the road, which still assumes that current levels of government support will continue for several decades.

5. PHEVs will have little impact on oil consumption before 2030 because there will not be enough of them in the fleet. More substantial reductions could be achieved by 2050. PHEV-10s will reduce oil consumption only slightly more than can be achieved by HEVs. A PHEV-10 is expected to use about 20 percent less gasoline than an equivalent HEV, saving about 70 gallons in 15,000 miles. Forty million PHEV-10s would save a total of about 0.2 million barrels of oil per day. The current light-duty vehicle fleet uses about 9 million barrels per day. PHEV-40s will consume about 55 percent less gasoline than equivalent HEVs, saving more than 200 gallons of gasoline per year per vehicle.

6. PHEV-10s will emit less carbon dioxide than non-hybrid vehicles, but save little relative to HEVs after accounting for emissions at the generating stations that supply the electric power. PHEV-40s are more effective than PHEV-10s, but the GHG benefits are small unless the grid is decarbonized with renewable energy, nuclear plants, or fossil fuel fired plants equipped with carbon capture and storage systems.

7. No major problems are likely to be encountered for several decades in supplying the power to charge PHEVs, as long as most vehicles are charged at night. Generation and transmission of electricity during off-peak hours should be adequate for many millions of PHEVs, although some distribution circuits may need upgrading if they are to serve clusters of PHEVs. Encouraging PHEV owners to charge their vehicles during off-peak hours will require both rate schedules that reward time-appropriate charging and equipment that can monitor—or even control—time of use.

8. A portfolio approach to research, development, demonstration, and, perhaps, market transition support is essential. It is not clear what technology or combination of technologies—batteries, hydrogen, or biofuels—will be most effective in reducing the nation's oil dependency to levels that may be necessary in the long run. It is clear, however, that a portfolio approach will enable the greatest reduction in oil use. Increasing the efficiency of conventional vehicles (including HEVs) beyond the current regulatory framework could reduce gasoline consumption by about 40 percent in 2050, compared to the Reference Case. Adding biofuels would reduce it another 20 percent. If PHEV-10s are also included at the Maximum Practical rate, gasoline consumption would be reduced an additional 7 percent, while PHEV-40s could reduce consumption by 23 percent. Employing HFCVs instead of PHEVs could eliminate gasoline use by the fleet.

1

Introduction

Light-duty vehicles play a crucial role in two of the key challenges facing the United States: energy security and climate change. Transportation is responsible for more than two-thirds of U.S. oil consumption, and about 60 percent of the oil we use must be imported. Dependence on imported oil leads to concerns over vulnerability to disruptions, especially if world oil production peaks. Burning that oil in vehicles also accounts for one-third of U.S. emissions of carbon dioxide (CO_2), the main greenhouse gas linked to global climate change. The U.S. government is seeking to reduce the use of oil to help meet both challenges. This report assesses the contributions that plug-in hybrid electric vehicles (PHEVs) can make to this effort.

The National Research Council (NRC) report *Transitions to Alternative Transportation Technologies—A Focus on Hydrogen* (2008) analyzed the potential for hydrogen-fueled fuel cell vehicles to penetrate the market and estimated the reductions in oil consumption and CO_2 emissions that might result. The report also compared these benefits to those that might be achieved by two alternatives: vehicles operating on biofuels and vehicles with advanced internal combustion engines. The latter included hybrid electric vehicles (HEVs) but not plug-in hybrid electric vehicles (PHEVs) or all-electric vehicles (EVs).

In 2009 the U.S. Department of Energy asked the NRC to extend its analysis to PHEVs, putting them on the same basis as the other alternatives to fuel cell vehicles. This report is the result of that additional task. The statement of task is in Appendix D.

PHEVs have recently been the focus of much attention, in large part because of rapidly improving battery technology. Several manufacturers intend to introduce PHEVs over the next few years. PHEVs are similar to today's HEVs, but they have larger batteries that can be charged from the electric grid and can supply sufficient energy to propel the vehicle for many miles. When PHEV batteries are discharged, the gasoline engine takes over, by either recharging the battery or directly providing power for propulsion. Short trips could avoid the use of gasoline altogether, and long trips are possible without the risk of being stranded, which is a concern for all-electric vehicles. PHEVs promise to reduce the use of gasoline without necessitating the major infrastructural changes that would be required for hydrogen, thus allowing an evolutionary transition from conventional vehicles. In addition, electric utilities may promote PHEVs because nighttime charging would help smooth out demand.

This report first evaluates battery and vehicle technologies to predict how costs might drop as technology improves and economies of scale increase. It considers PHEVs that can travel 10 (PHEV-10) and 40 (PHEV-40) miles on electric power as representative of all the PHEVs that may be available. Next it examines the ability of the electric grid to supply power for a growing PHEV fleet. Then it analyzes two potential market-penetration rates for PHEVs: (1) a Maximum Practical scenario, which makes optimistic assumptions about the evolution of PHEV technology, especially batteries, and about the barriers to market penetration and (2) a Probable scenario based on more likely assumptions. Because initially PHEVs will be considerably more expensive than equivalent conventional vehicles, the committee used the model developed in the hydrogen study to estimate the costs involved in supporting a transition to PHEVs. It also estimated the reduction of petroleum consumption and CO_2 emissions that could result from these two scenarios. Finally, the report discusses the committee's conclusions. The appendixes provide additional information on this study and the scenario analysis, plus a glossary of the many acronyms used in the report.

The committee was not able to analyze one potential barrier—the changes that PHEVs will require of drivers. The vehicles analyzed in the hydrogen report, even fuel cell vehicles, are functionally similar to current vehicles because they have the same range and refueling patterns. PHEVs, however, will require drivers to plug in their vehicles essentially every day. That will require a place where they can plug it in, preferably a garage or at least a car port, and the willingness to take the time to do it. While many people have the place, their willingness is a great uncertainty.

2

Batteries and Battery Packs for PHEVs

Battery cells, and the packs into which they are assembled, are the key component that will largely determine the viability of PHEVs. The battery packs must be affordable, durable, and safe. No commercially available battery meets all these requirements. Rechargeable lithium-ion (Li-ion) batteries, made by the billions for small electronic devices, are the most promising technology for automotive propulsion and will be used in the first-generation PHEVs soon to be rolled out. This section reviews the relevant technologies and estimates how their characteristics may evolve over the coming years.

TYPES OF PHEVS

Several configurations are possible for PHEV drive trains. The two considered here represent those that may be introduced, as shown in Figure 2.1. A PHEV differs from a hybrid electric vehicle (HEV) in that the battery can be charged from the electrical grid and operate the vehicle independently of the internal combustion engine (ICE) for a limited all-electric range (AER). This is the charge-depleting mode of operation. The ICE starts when the battery reaches its minimum state of charge and operates the generator to charge the battery. This charge-sustaining mode of operation prevents the battery from being discharged too deeply.

The PHEV-10 is designed for an AER of 10 miles before the ICE must start. It is similar to the Toyota Prius but has a larger battery and modified control electronics. Its split-power blended (or parallel-drive) configuration can drive the car either with only the electric motor powered by the battery or with the gasoline engine. When the battery is discharged to its minimum allowable level, the engine starts and the vehicle operates in a charge-sustaining mode, as in a conventional HEV. The engine will also start and assist in driving the wheels when more power is needed than can be delivered by the electric motor for rapid acceleration or heavy-load hill climbing. The PHEV-10 requires a more robust battery than an HEV because it must operate over a wide state of charge (SOC) range, enduring many deep charge/discharge cycles.[1]

The PHEV-40 has its engine, battery, and electric motor in series. The engine only charges the battery, and all propulsion comes from the electric motor. Thus it has a larger battery and motor than the PHEV-10 but a longer AER. It is conceptually similar to the General Motors Volt in design.

The size of the battery required to provide propulsion depends on the size and weight of the vehicle and the AER desired. For simplicity, this report considers just one vehicle, a midsize car, as representative of the fleet of light-duty vehicles, as was done in the 2008 Hydrogen Report. While midsize cars may not perfectly represent the fleet, they are adequate to illustrate the critical issues. Various recent studies have reported a range of energy requirements for midsize cars: in a study of Prius conversions to PHEVs, on average the vehicles required 238 watt-hours (Wh) per mile (Francfort, 2009). A simulated driving analysis calculated 170-200 Wh per mile energy consumption.[2] In addition, the GM Volt is expected to reach 40 miles on 8 kWh from its batteries, or 200 Wh per mile.[3] The committee assumed that the vehicles it analyzed would initially require 200 Wh per mile in its calculations. Larger, heavier vehicles would require substantially more energy per mile and bigger, more-expensive batteries, but those are not considered in this report. All vehicles are expected to become more efficient over time, and PHEVs will require less gasoline and less electricity, as discussed in Chapter 4.

[1]A rechargeable battery can be charged to 100 percent of its capacity and then discharged to 0 percent, but full charge would not allow regenerative braking, and full discharge typically would seriously damage its future performance. Early PHEV batteries may be limited to 80 percent of full charge and prevented from discharging to less than 30 percent. This is a 50 percent SOC range. Later generations may be able to operate with a wider range.

[2]M. Wang, A. Elgowainy, A. Burnham, and A. Rousseau, Center for Transportation Research, Argonne National Laboratory, Well-to-wheels energy and greenhouse gas results of plug-in hybrid electric vehicles, presentation to the committee, June 18, 2009, Washington, D.C.

[3]M. Verbrugge, General Motors, Extended-range electric vehicles, presentation to the committee, May 18, 2009, Washington, D.C.

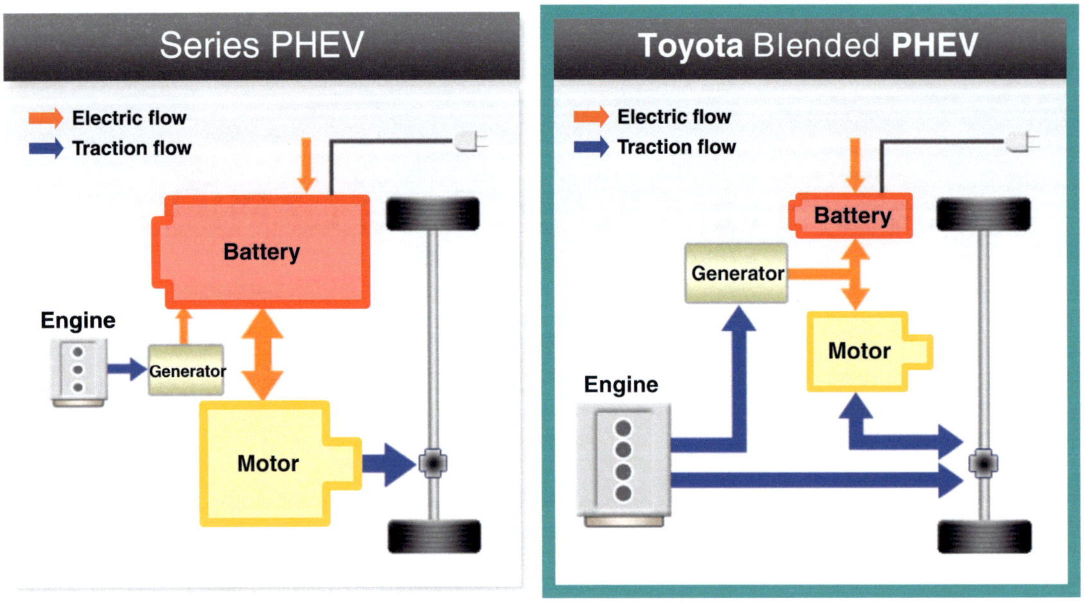

Mode	AER	Blended
Charge-depleting	Battery only	mainly Battery + Engine if needed
Charge-sustaining	mainly Engine	mainly Engine + Battery if needed

FIGURE 2.1 Plug-in hybrid electric vehicle concepts. SOURCE: Toyota.

Thus, the PHEV-10 requires 2.0 kWh of battery energy (actually used) to drive its 10-mile AER. The PHEV-40 draws 8 kWh of battery-stored energy to meet its 40-mile AER in charge-depletion mode before the engine starts and begins supplying power to operate the vehicle in charge-sustaining mode.[4] For the 50 percent SOC assumed in this report for the first generation of vehicles, the nameplate capacities are 4 kWh for the PHEV-10 and 16 kWh for the PHEV-40.

LITHIUM-ION BATTERY CELL CHEMISTRIES

For PHEVs to be widely accepted by consumers, batteries must be significantly cheaper than they are now, durable enough to have a long life, and safe. In addition, they will have to meet performance goals, which will require

- High power density to deliver the current needed for demanding driving conditions;
- High energy density for storing the needed energy for an extended all-electric range; and
- Wide range of SOC while maintaining a long cycle life.

[4]The batteries for these two vehicles are not identical because they are optimized for different conditions. For example, the PHEV-10 is likely to operate more in a charge, sustaining mode at minimum SOC than the PHEV-40.

Li-ion batteries currently are the only serious option for PHEVs. They are smaller and lighter than other batteries, and they promise to withstand multiple large SOC swings while maintaining their performance. They have more than twice the energy density and about three times the power density of the nickel-metal-hydride (NiMH) batteries used in current HEVs, and four times the energy density of the lead-acid batteries used in most vehicles today.

What no Li-ion battery can do—yet—is simultaneously deliver both high power density and high energy density at a reasonable cost. To meet this challenge, several promising Li-ion chemistries are being vigorously pursued by companies, research institutions, and governments. The technology is advancing rapidly, but there is no guarantee that any Li-ion battery will be developed that meets all goals for vehicle use. Table 2.1 compares the attributes of four of the more promising Li-ion battery chemistries.

Li-ion battery manufacturing technology is essentially the same for all battery chemistries. Typically the electrodes of Li-ion batteries are coated on metal foils, usually copper foil for the negative electrode and aluminum foil for the positive electrode, separated by an electrolyte (Nelson et al., 2009). Typical electrolytes are derived from solutions of $LiPF_6$ salt in a solvent blend of ethylene carbonate and various linear carbonates, such as dimethyl carbonate (Tikhonov and Koch, 2009; Zhang et al., 2002).

TABLE 2.1 Characteristics of Li-Ion Batteries Involving Different Chemistries

	Cathode/Anode			
Characteristics	Nickel Cobalt Aluminum Oxide/Graphite	Manganese Spinel/ Graphite	Iron Phosphate/ Graphite	Manganese Spinel/ Lithium Titanium Oxide
Durability	Good	Fair	Good	TBD
Power	Fair	Fair	Good	Good
Energy	Good	Good	Fair	Poor
Safety and abuse tolerance	Poor; safety concerns	Fair	Good	Good
Cell voltage	3.6	3.8	3.3	2.5
Some battery developers	Johnson Controls/Saft	LG Chem Ltd.	A123	EnerDel
Associated vehicle manufacturers	Toyota/Ford	GM	Daimler HEV buses	

NOTE: Cathode chemistries are frequently referred to as involving a spinel crystal structure. Actually there are no pure spinel structures present in Li-ion batteries; spinel-like would be more accurate.

While the power density (W/kg) of the cell is fixed by the surface area of the electrode foil, the energy density (Wh/kg) can be varied over a limited, but significant, range simply by increasing or decreasing coating thickness. HEV batteries, which require high power more than high energy storage, have thin electrode coatings. By contrast, electric vehicle (EV) batteries require high energy density and have thicker electrode coatings. Research has yielded new concepts for better electrodes and electrolytes. For example, raising the cell voltage to 5 V would increase the battery's energy density. Could this lead to a better PHEV battery? It is simply too early to tell.

In this report overall properties such as energy density, power density, and total energy available refer to the full range from 100 percent to 0 percent SOC. Energy and power density are intrinsic properties, and total available energy is the nameplate capacity of the cell or battery. For batteries in battery packs for vehicle operation, this report refers to the energy (kWh) actually used—that is, the nameplate capacity of all the cells in the pack multiplied by the allowable SOC.

LITHIUM-ION BATTERY PACKS

For PHEV applications, about 100 Li-ion cells are connected in series to provide the design voltage to operate the electrical propulsion motors. These cell groups are then installed in parallel, as needed, to provide the energy to drive the motor for the distance desired. Battery packs consist of these groups of cells, the supporting frame, electronic controls, and cooling systems to protect the cells. The current focus is on improving battery durability, safety, and cost competitiveness.

Battery Durability

Auto manufacturers have indicated that they intend to offer an 8-year warranty in 49 states and a 10-year warranty in California on PHEV battery systems as part of the drive-train warranty. Current commercial Li-ion batteries typically last 3 to 4 years, which is a function of both the number of charge/discharge cycles and calendar life (Howell, 2009). Some degradation is inevitable; for the purposes of this report, about 20 percent over the warranty period is assumed. If the PHEV-40 is expected to still have its required 8 kWh (actually used) of energy needed for an AER of 40 miles with the same 50 percent SOC range in 10 years, it could be sized to provide 10 kWh (actually used) energy initially. The other option is to assume that the SOC range is increased over time to account for battery degradation, which could be adjusted every year when the vehicle is brought in for servicing. This is the approach that the committee chose for the estimations that follow. If degradation is not too large or does not accelerate with larger SOC range, this should be satisfactory, but until demonstrated it remains a concern.[5]

Figure 2.2 compares the SOC variation for PHEV and HEV batteries. In a PHEV, batteries must undergo multiple large SOC range cycles without significant degradation. A 10-year life would require the batteries to undergo at least 2,000 cycles and still stay within the prescribed performance range. At present, this requires limiting the SOC range. The right-hand figure shows the much narrower (and less demanding) SOC range of an HEV.

This study assumes that SOC varies at most between 30 and 80 percent, or 50 percent of the total charge. The 30 percent lower limit is near the minimum and serves to maintain power and energy during charge-sustaining mode. The upper limit allows charging from regenerative braking while preventing overcharging and the resultant rapid battery degradation. A 50 percent range in SOC does, however, come at a price: The battery must have a nameplate capacity twice as high as the amount of energy actually needed and delivered to meet performance targets. In other words, the PHEV-40 will need a nameplate battery rating of 16 kWh to supply 8 kWh of the energy actually used for its 40 miles of charge-

[5]If after 10 years of operation the battery pack has lost 20 percent of its capacity (16 kWh down to 12.8 kWh for the PHEV-40), the SOC would have to be raised to 62.5 percent to maintain the required 8 kWh usable energy.

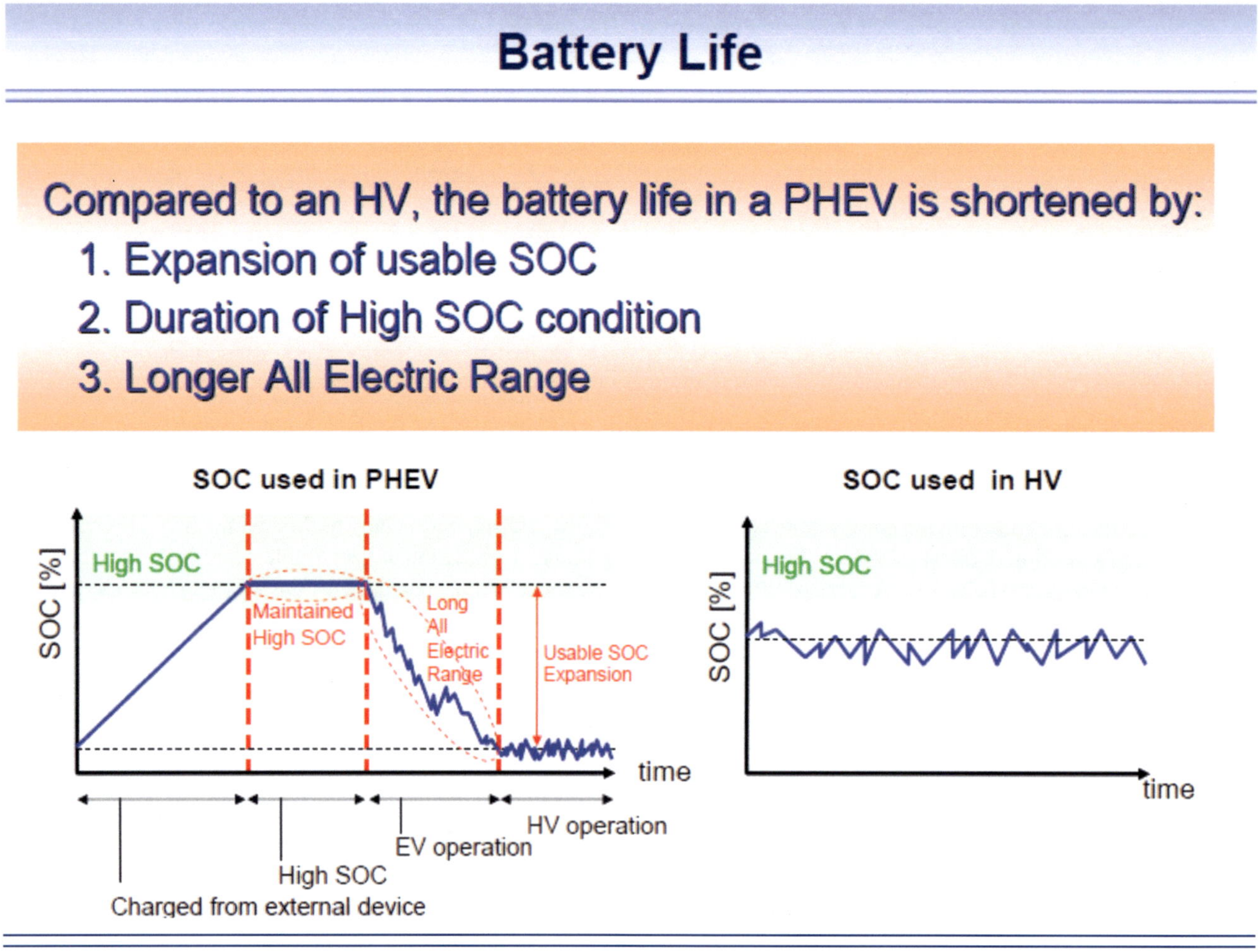

FIGURE 2.2 Differences in SOC requirements for PHEV batteries and HEV batteries. The PHEV is charged from an external source until it reaches its maximum state of charge, as shown on the left side of the figure. Its charge-depleting mode in AER takes it down to its minimum state of charge. The jagged portion of this curve is from regenerative braking, which partially recharges the battery. The level portion is charge-sustaining operation with the engine maintaining the battery charge around its lower SOC. The HEV also recharges from regenerative braking but operates in a much narrower SOC range. SOURCE: Toyota, presentation to the committee, May 18, 2009.

depleting driving (or 20 kWh if it is oversized to account for 20 percent degradation). Energy used is the product of the nameplate energy and the SOC range. Increasing the SOC range will increase the fraction of the nameplate energy used from a given battery pack size if that can be done without compromising durability.[6] Generally, however, the industry believes that for the first 5 years or so, battery durability issues will require conservative battery management—that is, keeping the SOC range at about 50 percent.

New technology may help meet the durability challenge. One type of battery is claimed to have lasted 7,000 cycles in accelerated aging tests covering a wide (70 to 90 percent) range in SOC with little reduction in performance under controlled test conditions.[7] However, life prediction is difficult, and actual performance will not be known until many vehicles are in service for many years. In the absence of this operational information, accelerated-age tests are used to estimate the expected performance, but they may not capture the full effect of actual aging.

In addition, higher temperatures and other excursions outside the design envelope (e.g., SOC limits and rate of charging) detract from durability and battery life. Accordingly, cooling and temperature control systems will have to be included in the battery pack, and operation-control strategies must avoid excursions in operational performance.

[6]While increased SOC range is one of the factors leading to cost reductions in the required battery pack (discussed later), it can also increase the rate of degradation.

[7]D. Vieau, A123 Systems, Lithium-ion battery progress, presentation to the committee, May 2009, Washington, D.C.

Battery-Pack Safety

Safety has been a concern with Li-ion batteries, which can overheat and catch fire or even explode, emitting burning gases. There appear to be two separate causes for these thermal runaways: contaminants and overcharging.

Contaminants, particularly small metal particles, can enter the cell during manufacturing, causing a short circuit between the anode and the cathode, resulting in a fire. Improved manufacturing techniques and rigorous quality control should manage this issue, albeit at an increased cost. Overcharging the cells or charging them too rapidly can lead to overheating, which can degrade the battery and limit its service life. With some Li-ion chemistries, overcharging can result in thermal runaway and catastrophic failure. Control systems to prevent this are discussed in the following section.

Proper choice of Li-ion chemistry, controlled manufacturing procedures, and onboard monitoring and temperature control should assure safe batteries and safe battery pack operation. Another safety concern, crashes, also must be resolved. Passengers and emergency workers must be safe from shocks and fumes. As with durability, battery and automotive manufacturers are confident that the safety issues can be overcome and managed. The consequences of catastrophic failures would be too great for manufacturers to market PHEVs that do not meet very high safety standards.

Battery Pack Cooling and Control Electronics

The battery pack, in addition to containing a hundred or so interconnected Li-ion cells, includes two control systems essential for the safety and durability of the batteries. Both of these systems include significant electronics and other equipment, located separate from, but connected to, the battery pack.

One of these systems monitors and controls the temperature of the battery cells. With the current state of Li-ion battery cell technology, the individual cell temperature should not exceed 60°C (140°F) because the batteries deteriorate at higher temperature.[8] The electrically driven temperature-control unit uses cooling fluid to maintain battery temperature. Liquid cooling is assumed to be required for larger battery packs; smaller battery packs, such as for a PHEV-10, may allow air cooling.

The other system measures the voltage of each cell and ensures that it does not exceed an upper limit during charging or regenerative braking, which could lead to thermal runaway (overheating). The voltage of each Li-ion cell should not exceed its specified value by more than a few tens of millivolts. This balancing of charge is important to battery life and battery safety.

The non-cell portion of the battery pack (i.e., the structure and control systems) could account for around 50 percent of the pack's cost and is less likely to produce large future cost reductions. However, with improved battery technology, operational experience, and better quality control in cell manufacture, it might be possible to monitor some of the cells rather than all of them, as is done now, which would help reduce battery pack cost. In addition to adding to costs, cooling and monitoring of the cells add significantly to the weight of the battery pack, reducing the power and energy density of the battery pack.

Battery-Pack Performance and Cost

Tables 2.2 and 2.3 summarize the committee's estimates of Li-ion battery and battery-pack performance, and costs for the two PHEV types examined in three time periods, 2010 (current technology), 2020, and 2030. Additional detail on the committee's analysis of battery-pack cost can be found in Appendix F. These estimates were arrived at after literature searches and discussions with industry experts. The values in the tables represent the judgment of the committee of the most probable rate of anticipated progress based on the entirety of the data available to it.[9] Future battery and battery-pack costs are quite uncertain at this point. For that reason the committee feels that it will be important to reevaluate these costs in several years, when significant data on the first production cycle of PHEVs is available, which should allow better projections.

Optimistic and conservative estimates also were made for the production costs expected in 2010, 2020, and 2030. "Optimistic" means progress is faster than expected. "Conservative" means partial rather than "probable" success but could also mean that additional battery capacity (and thus cost) was necessary to account for degradation. The "probable" and "optimistic" estimates form the basis for the projections that the committee modeled, as discussed in the next section. Box 2.1 lists DOE's battery targets for comparison.[10] It is the committee's opinion that these PHEV battery goals are extremely aggressive and are unlikely to be reached by the target date or even for a significant time beyond.

---

[8]Damage can start at 50°C, but deterioration is slow. Vehicles are unlikely to be exposed to such high ambient temperatures, but the heat given off by charging and discharging can lead to high temperatures inside the pack. Thus cooling is necessary, particularly in hot climates.

[9]The performance and cost numbers in Tables 2.2 and 2.3 are less optimistic than some others that have been claimed. Lithium-ion battery manufacturing is a well-developed technology. Worldwide over a billion Li-ion cells are currently produced every year. They are made by coating large sheets that are then cut up in small pieces for cell phones and other electronic devices. Vehicle batteries will be conceptually similar, but the sheets will be cut in larger pieces. Also, a large part of the cost of automotive batteries is the packaging, which involves electronics for monitoring the cell voltage and state of charge the SOC, cooling systems and, their mechanical supports and sheet metal. These components are not expected to decline greatly in cost.

[10]D. Howell, DOE, DOE targets for battery performance, presentation to the committee, June 2009.

TABLE 2.2 Estimates of Li-Ion Battery Performance Parameters for a PHEV-40

Characteristic		2010[a]	2020	2030
Energy density at nameplate cell level, Wh/kg	Probable	150	200	200
Power density at nameplate cell level, W/kg for 12 sec	Probable	1,400	1,600	1,750
Energy density at nameplate battery pack level,[b] Wh/kg	Probable	120	150	150
Power density at nameplate battery pack level,[c] W/kg for 12 sec	Probable	1,150	1,250	1,400
Cycle life over SOC at 40°C ambient	Probable	3,000	5,000	7,500
Battery pack cost per kWh over SOC variation (8 kWh actually used), $/kWh[d]	Conservative	2,000	1,275	1,150
	Probable	1,750	1,120	1,000
	Optimistic	1,250	800	720
Battery pack cost per kWh for nameplate energy level (16 kWh), $/kWh	Conservative	1,000	638	575
	Probable	875	560	500
	Optimistic	625	400	360
Battery calendar life, yr	Conservative	3	7	9
	Probable	5	10	10
	Optimistic	8	12	15

NOTE: PHEV-40 nameplate battery rating 16 kWh (8 kWh usable); SOC variation range, 80-30 percent; 100+ kW peak power.

[a]First production cycle.

[b]Battery pack means the entire system, including packaging, cooling, and monitoring and control electronics.

[c]Power density numbers for PHEVs are still variable since developers are engineering their cells to give optimum life and energy.

[d]As applied to SOC range actually used. Cost per kWh based on nameplate capacity would be half these. Additional information on the committee's analysis of these costs is in Appendix F.

TABLE 2.3 Estimated Battery Performance Properties for a PHEV-10

Characteristic		2010[a]	2020	2030
Energy density at nameplate cell level, Wh/kg	Probable	100	150	150
Power density at nameplate cell level, W/kg for 12 sec	Probable	1,500	1,600	1,750
Energy density at nameplate battery pack level, Wh/kg[b]	Probable	80	110	125
Power density at nameplate battery pack level,[c] W/kg for 12 sec	Probable	1,250	1,350	1,400
Cycle life over SOC at 40°C ambient	Probable	3,000	5,000	7,500
Battery pack cost per kWh over SOC variation (2 kWh actually used), $/kWh[d]	Conservative	2500	1,600	1,450
	Probable	1,650	1,050	950
	Optimistic	1,250	800	725
Battery pack cost per kWh for nameplate energy level (4 kWh), $/kWh	Conservative	1,250	800	725
	Probable	825	525	475
	Optimistic	625	400	363
Battery calendar life, yr	Conservative	3	7	9
	Probable	5	10	10
	Optimistic	8	10	15

NOTE: PHEV-10 nameplate battery rating 4.0 kWh (2 kWh usable); SOC variation range, 80-30 percent; 50+ kW peak power.

[a]First production cycle.

[b]Battery pack means the entire system, including packaging, cooling, and monitoring and control electronics.

[c]Power density numbers for PHEVs are still highly variable since developers are engineering their cells to give optimum life and energy.

[d]As applied to SOC range actually used. Cost per kWh based on nameplate capacity would be half these. Additional information on the committee's analysis of these costs is in Appendix F.

BOX 2.1
Department of Energy Targets for Battery Performance

The Department of Energy has set several targets for battery performance:

- *Battery cost.*
  - — PHEV-10: (3.4 kWh available energy at end of life)[1]
    $500/kWh or $1,700 battery cost[2] achieved in 2012 vs. $1,000+/kWh today
  - — PHEV-40: (11.6 kWh available energy at end of life)
    $300/kWh or $3,400 battery cost[2] achieved in 2014 vs. $1,000+/kWh today
- *Battery life.*
  - — PHEV-10: 10+ years achieved in 2012 (5,000 cycles) vs. 3+ years today
  - — PHEV-40: 10+ years achieved in 2014 (3,000-5,000 cycles)
- *Maximum system weight.*
  - — PHEV-10: 60 kg for in 2012 vs. 80-120 kg today
  - — PHEV-40: 120 kg in 2014

[1]This PHEV-10 is a small SUV that requires more energy than the midsize car modeled in this study.
[2]At high volume production.
SOURCE: Adapted from DOE (2009b).

PROJECTED PHEV INCREMENTAL COSTS

Tables 2.4 and 2.5 compare the current incremental cost of components for a PHEV-40 and a PHEV-10 with those of a conventional (nonhybrid) car. Savings from eliminating components or reducing size are shown as negative numbers; for example, the automatic transmission can be eliminated when the drive is electric.[11] These incremental numbers are for the first round of PHEV production, including the estimated cost of the battery pack, the least well defined of the costs. Initially, the PHEV-40 is expected to cost the vehicle manufacturer about $18,000 more than an equivalent conventional car and the PHEV-10 to cost about $6,300 more. The price to the customer, before government subsidies, is likely to be significantly higher once manufacturers' additional expenses and profit and dealers' markup are added in.

These costs are likely to decline over time. Table 2.6 summarizes projections of cost reductions for the different components for the two PHEV types for 2015, 2020, and 2030. Reduction estimates are posited on technology improvements, on experience gained over time through several cycles of technology evolution, and from increased economies of scale. The committee held discussions with various experts who provided valuable input to this table. There was good agreement on the expected rate of improvements, particularly for the non-battery components, where there is considerable experience. There also was general agreement that battery pack costs would decline significantly, but not dramatically, for the first 10-15 years of commercial experience and would later slow.

The reductions expected mirror the experience with NiMH batteries for HEVs, where costs came down significantly at first but then decreased much more slowly. The NiMH battery pack for HEVs saw a cost reduction of about 11 percent from 2000 to 2006 but since has seen much less change. Li-ion battery cost decreased by about 35 percent from 2000 to 2008, but most of that was at the beginning of that period, with only about 5 percent after 2004 (Howell, 2009). Manufacturers of Li-ion batteries with technology similar to consumer batteries are already considerably further along the learning curve than were manufacturers of NiMH batteries when HEVs were introduced, so steep cost reductions seem unlikely. Nor does it seem likely that the cost of materials will decline greatly. Indeed, some materials, including lithium, may increase in cost with additional demand, but the committee believes that the supply of lithium will be adequate for any plausible number of PHEVs manufactured worldwide.

It is likely that much of the reductions in Li-ion cell costs will come from technology innovations, with smaller reductions from manufacturing improvements and volume

[11]The PHEV-10 will require a transmission because the engine is connected directly to the wheels. However, the committee assumed that manufacturers would use a small, electronically controlled continuously variable transmission (ECVT) such as used in the 2010 Prius. This cost is included under power electronics in Table 2.5.

TABLE 2.4 Projected Incremental Cost[a] of Components for PHEV-40 for Production in 2010 Using Current Technology Compared with an Equivalent Current Nonhybrid Vehicle

Component		Price That a Supplier Charges the Vehicle Manufacturer for the Technology	Cost Reductions in Components due to Vehicle Changes in Going to PHEV-40	Incremental Cost of PHEV-40 Vehicle vs. Modern, Comparable ICE Vehicle
Motor/generator	Probable	1,800		1,800
Power electronics, DC/DC converter (1.2 kW), and inverter	Probable	2,500		2,500
Li-ion battery pack	Conservative	16,000		16,000
8 kWh actually used	Probable	14,000		14,000
16 kWh nameplate capacity[b]	Optimistic	10,000		10,000
Electrical accessories	Probable	100		100
Electric air conditioning	Probable	400		400
Regenerative brakes	Probable	180		180
Electric power steering/water pump	Probable	200		200
Body/chassis/special components	Probable	200		200
Automatic transmission	Probable		850	−850
Starter and alternator	Probable		95	−95
Engine simplification	Probable		300	−300
Total	Conservative	21,380		20,135
	Probable	19,380	1,245	18,135
	Optimistic	15,380		14,135

[a]Series plug-in hybrid 40-mile AER, 100+ kW peak power, 8 kWh usable; 16 kWh nameplate capacity.
[b]See Appendix F for further information on the committee's analysis of costs.

TABLE 2.5 Projected Incremental Cost[a] of Components for PHEV-10 for Production in 2010 Using Current Technology Compared with an Equivalent Current Nonhybrid Vehicle

Component		Price That a Supplier Charges the Vehicle Manufacturer for the Technology	Cost Reductions in Components due to Vehicle Changes in Going to PHEV-10	Incremental Cost of PHEV-10 Vehicle vs. Modern, Comparable ICE Vehicle
Motor/generator	Probable	1,500		1,500
Power electronics, DC/DC converter (1.2 kW), and inverter	Probable	1,500		1,500
Li-ion battery pack	Conservative	4,000		4,000
2.0 kWh actually used	Probable	3,300		3,300
(4 kWh nameplate capacity)[b]	Optimistic	2,500		2,500
Electrical accessories	Probable	100		100
Electrical air conditioning	Probable	400		400
Regenerative brakes	Probable	180		180
Electric power steering and water pump	Probable	200		200
Body/chassis/special parts	Probable	200		200
Automatic transmission	Probable		850	−850
Starter and alternator	Probable		95	−95
Engine simplification	Probable		120	−120
Total	Conservative	8,080		7,015
	Probable	7,380	1,065	6,315
	Optimistic	6,580		5,515

[a]Split-power plug-in hybrid, 10-mile AER capacity, 50+ kW peak power, 2 kWh usable; 4 kWh nameplate capacity.
[b]See Appendix F for further information on the committee's analysis of costs.

TABLE 2.6 Percent Projected Cost Reductions for Different Components with Increased Production and Learning by Doing

Component	Year Reduction Achieved/ Year Against Which Compared		
	2015[a]/ 2010	2020[b]/ 2015[a]	2030[c]/ 2020[b]
Motor/generator/gear set	5	5	5
Power electronics, AC/DC converter	10	15	5
Li-ion battery pack	25	15	10
Electrical accessories	5	5	5
Air conditioning	10	5	5
Regenerative brakes	5	5	5
Electric power steering + water pump	5	5	5
Body/chassis/special components	10	5	5

NOTE: Estimated cost reductions are based on increased production volumes and anticipated improvements in technology and production techniques. Unanticipated technology advances (breakthroughs) could lead to faster reductions.

[a]Assumed production, 25,000 vehicles per year.

[b]Assumed production, 1 million vehicles per year.

[c]Assumed production, 1 million-plus vehicles per year.

production.[12] Although it is hard to quantify, about half of the cell cost is estimated to be for materials, and the cells account for about half the battery pack cost, further reducing the impact of cell-only cost reductions.

The additional costs for changes in mechanical and electrical components in going from a conventional vehicle to a PHEV are considered quite predictable and have the expected impact on vehicle cost. These estimates (Table 2.4 and Table 2.5) are only for the cost of the components to the vehicle manufacturer and do not include the cost for vehicle engineering, R&D, or the automakers' capital investments. These and other markups to the vehicle price, which is what the customer will see, are addressed in Chapter 4 of this report.

Overall, Li-ion battery-pack costs may decline by almost 50 percent, as shown in Table 2.2, from $1,750 per kWh energy actually used in 2010 to about $1,000 per kWh in 2030. Collectively, the reductions in component costs lead to future PHEV costs shown in Table 2.7. These estimates do not consider the possibility of technological breakthroughs, which, if they occur, could significantly reduce the costs and improve the viability of PHEVs. Table 2.7 and the scenarios that follow do not report the conservative estimates, for if costs remain that high, PHEVs are unlikely to achieve much success in the market.

[12]D. Vieau, A123 Systems, Lithium-ion battery progress, presentation to the committee, May 2009, Washington, D.C.

TABLE 2.7 Estimated PHEV Incremental Costs

	2011[a]	2015	2020	2030
PHEV-40	14,100-18,100	11,200-14,200	9,600-12,200	8,800-11,000
PHEV-10	5,500-6,300	4,600-5,200	4,100-4,500	3,700-4,100

NOTE: These are the incremental costs to manufacture the vehicle itself, relative to a conventional (nonhybrid) vehicle. They do not include engineering, overhead, or other costs, or profit, and thus are not the total incremental prices to the customer. Ranges represent probable and optimistic assessments of battery technology progress.

[a]Costs for 2011 are based on low battery production rates in response to contracts initiated about 2 years earlier.

OTHER TECHNOLOGY OPTIONS AND POTENTIAL BREAKTHROUGHS

The cost of Li-ion batteries is currently very high, making it difficult for PHEVs to be cost competitive when the cost of gasoline is less than $4 per gallon. Although considerable progress is expected in reducing battery costs, it is not clear that sufficient cost reductions can be achieved with Li-ion batteries or battery packs to make PHEVs cost competitive without substantial subsidies.

Announcements continue from researchers about improvements in Li-ion batteries, including better electrodes and electrolytes and, possibly, higher cell voltages (to 5 V), resulting in better energy density. Unfortunately, it is hard to evaluate the practicality of these concepts or to assess which, if any, will become commercial and when.

Other Li-ion battery cell chemistries may offer better performance than those currently projected for PHEV applications,[13] but serious questions remain about their durability, safety, and costs. There appears to be little chance that any of these could become commercially cost competitive in the near future.

A breakthrough in battery technology would definitely improve the prospect of PHEVs becoming economically competitive. It is not possible to predict or schedule scientific and technical breakthroughs, but a continued, substantial scientific research effort is needed to increase the chances that this will occur. However, even if a breakthrough occurs, it will be decades before it has a great impact. Major battery developments will require considerable work and time prior to commercialization to confirm cost advantage, durability, and safety, and years more to achieve significant penetration into the fleet.

Options such as the lithium-air battery and solid polymer Li-ion electrolyte batteries are under study. Several large U.S. corporations are working on lithium-air technology, which could offer 5 to 10 times as much energy density as the Li-ion batteries discussed above. This battery is much

[13]D. Vieau, A123 Systems, Lithium-ion battery progress, presentation to the committee, May 2009, Washington, D.C.

lighter, but there are issues of safety, primarily because of lithium's reactivity with water, and regeneration or recharging needs to be developed. Solid polymer Li-ion electrolyte batteries offer higher energy densities and more stability than Li-ion batteries, but safety and operational challenges (such as achieving acceptable current density at ambient temperatures) will be difficult to meet. There do not appear to be any other radically new battery technologies on the horizon (the lithium metal-air battery concept has been around for many years) that could economically provide the enhanced performance needed, but the vibrant research and development programs world-wide may produce a technology that will overcome these barriers.

Also, totally different approaches are being considered. Swapping battery packs at stations that charge them for the next vehicle is one possibility, but it is not clear if pack and vehicle design will be sufficiently standardized to make this widely practical. Battery leasing is another proposal. Leasing could lower the initial cost to the consumer and perhaps provide some reassurance about durability, but it would not necessarily lower overall costs.

It should also be noted that higher CAFE standards or high oil prices will improve the competitiveness of PHEVs. Conversely, HEV cost and performance characteristics will continue to improve, reducing the fuel-saving advantage of PHEVs. Although HEVs will be more expensive than nonhybrid vehicles, PHEVs will be significantly more expensive than HEVs. However, the low fuel consumption of PHEVs, especially the PHEV-40 type, will be advantageous in helping the United States reduce its dependence on imported oil. Also, once the carbon intensity of grid electricity is reduced, PHEVs will be able to significantly reduce greenhouse gas (GHG) emissions from the light-duty vehicle sector.

3

U.S. Electric Power Infrastructure

PHEVs require electric power to charge their onboard batteries. Unlike fuel cell vehicles, which would require a brand new supply infrastructure, PHEVs have a ready and well-established energy source—the U.S. electric power system. This vast system includes a variety of fuel sources and generation technologies, a nationwide transmission network, and distribution operations that reach almost all Americans.

This chapter begins with a brief overview of the current system. It then describes two projections of how the system might evolve by 2050, one based on current policy and the other representing a concerted effort to reduce U.S. CO_2 emissions. This section also discusses the charging of PHEVs and its potential impact on the electric system. Finally, it introduces several issues that are relevant but beyond the scope of this study.

U.S. ELECTRIC POWER SYSTEM

The nation's 1 million megawatts (MW) of electric generating capacity produced over 4 billion megawatt-hours (MWh) in 2007 (EIA, 2009b). In comparison, 1 million PHEVs charging an average of 3 kWh every day for a year would consume only about 1 million MWh.[1] The U.S. electric system can clearly handle a great many PHEVs, but there is one caveat. Electricity demand varies throughout the day and over the year. Demand usually peaks on hot afternoons when summer air conditioning loads are highest. On such days, some systems are seriously stressed—sometimes to the point where they have to shed loads (reduce demand) to avoid collapse.

In recent years, the North American Electric Reliability Corporation (NERC) has raised concerns about the reliability and development of the electric power system. In its 2007 report, NERC noted that "projected increases in peak demands continue to exceed projected committed resources beyond the first few years of the 10-year planning horizon" (NERC, 2007). In its 2008 report, NERC said that "while some progress has been made, action is still needed on all of the issues identified in last year's report to ensure a reliable bulk electric system for the future" (NERC, 2008).

Charging a large number of PHEVs during peak hours could aggravate a potentially serious problem, possibly increasing the risk of brownouts and other power system disruptions that could adversely impact the public's interest in PHEVs. Currently, electric system capacity is generally adequate, but as the economy recovers, demand will increase, stressing the system unless new generating and transmission capacity is built.

At the outset, the key to integrating PHEVs will be to encourage off-peak charging. Generation and transmission capacity must be adequate to handle peak loads, but most of the time, demand is much lower. Utilities would greatly prefer that PHEVs be charged at night, when they can employ their otherwise underutilized capacity or purchase power at lower rates. Many utilities offer time-of-use (TOU) rate structures to at least some of their residential customers, with lower rates at night than during peak hours.

Many plug-in hybrids can be charged with available power generation and grid capacity during off-peak hours. An analysis by the Pacific Northwest National Laboratory estimated that a PHEV fleet equal in size to 84 percent of all cars and light trucks on the road in 2001 could be charged during off-peak times without building new electric generation capacity (PNNL, 2007).

The picture is different if PHEVs are charged during peak hours. For example, a study by Southern California Edison concluded that PHEVs could account for as much as 11 percent of its system load by 2020, which could increase peak loads by several thousand megawatts if PHEV charging is not properly managed.[2]

[1]As analyzed in this report, a PHEV-10 has usable storage capacity of about 2 kWh, and a PHEV-40 has about 8 kWh.

[2]D. Cromie and B. Graham, Transition to electricity as the fuel of choice, Southern California Edison, presentation to the committee, May 2009.

The committee assumed that most PHEV charging will be accomplished at night, when electric power demand is lower and rates are likely to be lower than during the day. Encouraging PHEV owners to charge their vehicles during off-peak hours will require both rate schedules that reward time-appropriate charging and equipment that can monitor—or even control—time of use. Under the Energy Policy Act of 2005, utilities are "required to move towards smart meters that allow time-of-day pricing," and smart meters are already being installed in certain areas to improve electric service, encourage efficiency, and shift energy use to off-peak hours. Many utilities are planning to deploy smart meters within the next few years.

Modernizing the transmission grid to achieve a smart grid as well as distribution systems would also benefit PHEVs by improving reliability, accommodating daytime charging, helping reduce carbon emissions, and controlling costs (NAS-NAE-NRC, 2009). DOE recently released a solicitation offering $3.9 billion in grants to "modernize the electric grid, allowing for greater integration of renewable energy sources while increasing the reliability, efficiency and security of the nation's transmission and distribution system" (DOE, 2009a).

In its scenario analysis, the committee examined two cases that bracket the national average residential rate of 10.4 cents per kWh (EIA, 2009a) and that represent likely PHEV charging rates: 8 cents per kWh and 15 cents per kWh. The former would apply in areas with residential TOU rate structures; the latter would be in areas where rates are high or if they rise, perhaps because electric power generation is decarbonized.

CO_2 will be emitted from power plants that generate the electricity that replaces gasoline that PHEVs do not require relative to conventional vehicles. As shown in Figure 3.1, the primary sources of electric power in 2007 were coal, natural gas, and nuclear energy. From 1997 through 2007, these three sources provided between 84.6 and 89.5 percent of total net generation. Nuclear power generation releases no CO_2, but coal and (to a lesser extent) natural gas do.[3]

CO_2 emissions by U.S. electric generators and combined heat and power facilities in 2007 were 2,517 million metric tons (EIA, 2009b), or an average of about 1.3 pounds of CO_2 per kWh. One kWh will take a small electrically driven car about 5 miles. Over the same distance, an equivalent gasoline-powered car that gets 30 miles per gallon (mpg) would emit 3 pounds of CO_2, more than twice as much. An HEV at 50 mpg would release about 2 pounds.

[3]Some CO_2 is released from the nuclear fuel cycle, but the amount per kWh generated is small relative to fossil-fired power plants.

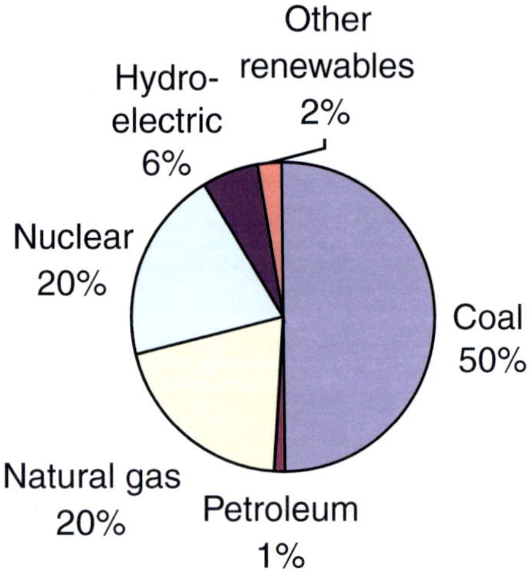

FIGURE 3.1 Net generation of U.S. electric power industry, 2007. SOURCE: EIA, 2009b.

THE SYSTEM OUT TO 2030 AND BEYOND

Energy Information Administration Projection (Business as Usual)

From 2000 to 2007, average electricity demand increased by 1.1 percent per year. The 2009 EIA Reference Case projects electricity demand increasing by 26 percent from 2007 to 2030—about 1.0 percent per year on average. The largest increase is in the commercial sector (38 percent), where service industries continue to lead demand growth, followed by the residential sector (20 percent) and the industrial sector (7 percent) (EIA, 2009a). EIA also provides low and high growth cases for 2030. Figure 3.2 compares the generation mix for the three cases in 2030 with the 2007 case.

EIA's Reference Case projects that the average retail price for electricity in 2030 will be very close that of 2008, 10.4 cents per kwh, with the high growth case at 10.8 cents and the low growth at 9.7 cents per kwh. These modest price differences are unlikely to have a material influence on PHEV economics and acceptance.

It should be noted that EIA forecasts are required to assume the continuation of existing policy, so no substantial efforts to reduce CO_2 emissions from electric generation were included. The committee used the EIA projections for its business-as-usual scenario.

An Alternative View: EPRI/NRDC (Policy Driven)

For PHEVs to deliver their full potential to reduce CO_2 emissions, the electricity used for charging them must be generated from technologies such as nuclear, renewable energy (e.g., solar, wind), and fossil fuels with carbon capture and sequestration. Because government policies will be required to drive these changes, the rate at which the country

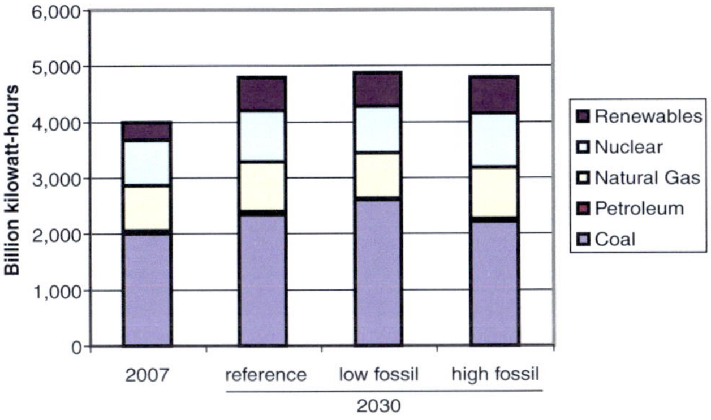

FIGURE 3.2 Electric generation by fuel in four cases: 2007 and 2030 (Reference Case, high growth, low growth). SOURCE: EIA, 2009a.

moves toward this greener power generation mix remains uncertain.

An alternative set of scenarios for U.S. power generation was developed jointly by the Electric Power Research Institute (EPRI) and the Natural Resources Defense Council (NRDC) to explore the relationship between the grid and PHEVs if it becomes necessary to lower CO_2 emissions from U.S. electric power generation (EPRI/NRDC, 2007). Nine modeling scenarios were developed spanning high, medium, and low emissions of CO_2 and low, medium, and high penetrations of the fleet by PHEVs. Chapter 4 compares greenhouse gas (GHG) emission intensities of the EIA Reference Case with the EPRI/NRDC medium case.

Among other things, EPRI and NRDC concluded that all nine cases showed significant GHG reductions attributable to PHEV fleet penetration. Cumulative GHG savings from 2010 to 2050 could be significant, ranging from 3.4 to 10.3 billion MT of CO_2.[4]

Recognizing that reductions of this magnitude are not likely to occur without public policy intervention, the committee used the EPRI/NRDC results to illustrate the potential benefits that PHEVs might provide under a policy-driven low-emission grid scenario.

CHARGING THE BATTERIES

If a dedicated circuit is not required, many PHEVs can be charged with little or no change to an owner's electrical service. Although significant upgrades in the electrical distribution system might be required for a large PHEV population, utility planners should have sufficient time to prepare for these changes.

Home Charging

Charging a PHEV may be a simple matter of finding a suitable electrical outlet (most likely in a home garage) and plugging in. In other cases, however, it will be more complicated. The time required to charge a PHEV at regular household voltage may be quite long, so a voltage upgrade may be necessary. Zoning codes or standards may require upgraded or dedicated service for PHEVs, and PHEV-friendly, off-peak charging may require the installation of dedicated charging circuits and/or meters.

One recent study considered three levels for PHEV charging (Morrow et al., 2008):

- Level 1 charging uses a standard 110 volt, 15 to 20 ampere circuit, standard in residential and commercial buildings. Level 1 provides relatively little power and may necessitate prolonged charge times.
- Level 2 charging involves a 220 volt, single-phase, 40 ampere circuit. At the higher voltages and currents, charging would be more rapid, but Level 2 service is not common in residential garages and would generally entail a system upgrade.
- Level 3 charging uses a 440 volt, three-phase circuit supplying 60-150 kW of power and can deliver a 50 percent charge in 10-15 minutes, depending on vehicle size and electrical range. Level 3 charging might be the choice for public garages, parking lots, and shopping centers.

The committee has considered charging only at Levels 1 and 2, believing that charging at Level 3 will not become important until much later. Table 3.1 provides estimated charging times for representative PHEVs and charging stations. Costs per charging station were estimated (numbers rounded by the committee) as follows (Morrow et al., 2008):

- Residential garage charging
  —Level 1, $880
  —Level 2, $2,100
- Apartment complex charging
  —Level 1, $830
  —Level 2, $1,500
- Commercial facility charging
  —Level 2, $1,900

At the time this report was prepared, manufacturers had not announced whether they would equip PHEVs for charging at both 110 and 220 volts. The committee believes, however, that the additional cost for dual voltage vehicle charging is probably small and not likely to significantly affect the committee's analysis.

In summary, some PHEV owners may be able to charge their vehicles using their existing home electrical service, but many others probably will not. The cost of upgrading

[4]Total CO_2 emissions from gasoline in the transportation sector currently equal about 1.2 billion metric tons per year.

TABLE 3.1 Approximate Charging Time as a Function of Vehicle Size and Electric Driving Range (hours)

	PHEV-10	PHEV-20	PHEV-40
Level 1			
Economy vehicle	2.7	5.5	11
Midsize vehicle	3.6	7.3	14
Light-duty truck/SUV	4.5	9.1	18
Level 2			
Economy vehicle	0.5	1	2
Midsize vehicle	0.7	1.3	2.7
Light-duty truck/SUV	0.8	1.7	3.3

NOTE: Numbers rounded by the committee.
SOURCE: Morrow et al., 2008.

home service to allow PHEV charging, whether desired or required, is estimated to range from slightly less than $1,000 to slightly more than $2,000. PHEV-40s are more likely to need costly new circuitry for 220 volts.

PHEV subsidies may soften the financial concerns associated with this issue, and in some cases (especially for meter upgrades), utilities may pay for such upgrades and amortize the costs over a series of electric bills. However, an open question remains: To what extent will these additional costs, or just the inconvenience of making the modifications, dissuade potential PHEV buyers?

Public Charging

As PHEVs proliferate, there will be a growing demand for public charging, much of which could occur during daytime hours, when electric power costs are higher. It seems likely that some office complexes will install chargers for their employees and visitors, and shopping malls may install chargers to attract customers. In some cases, businesses may not even charge for the electric power, treating it instead as a promotional expense.

As an indication of interest in public charging, one company, Electric Transportation Engineering Corp., was recently awarded a stimulus grant of nearly $100 million from the Department of Energy to build 12,800 charging stations for electric vehicles and PHEVs in Arizona, Washington, Oregon, California, and Tennessee (DOE, 2009b).

ADDITIONAL ISSUES

The committee identified some related issues that are beyond the scope of this study and will require detailed assessment to understand the impact of PHEVs on the grid and vice versa:

- *Outlet access.* An accurate estimate is needed of the number of existing homes and buildings where charging would be easy. About 35 percent of housing units do not have a garage or carport, which is probably essential for an outlet for home charging (Bureau of the Census, 2008). PHEV owners without ready access to an outlet would need a public charging infrastructure; it is uncertain how many consumers would be willing to rely on public charging.
- *Charging at 440 V.* Some carmakers may be interested in 440-V charging to reduce charging times (Carney, 2009). The cost and potential extent of such service needs study.
- *Distribution system upgrades.* In some areas, local utility electric distribution capacity may not be adequate for the simultaneous charging of many PHEVs on one circuit, particularly for fast charges. These areas should be identified and plans for upgrading developed.
- *Safety.* Safety issues associated with charging PHEVs must be thoroughly studied and problems minimized.
- *Energy stored in PHEVs.* It has been suggested that the electric grid might use the electric energy stored in PHEVs to help meet peak demand (when the costs of producing power are very high) and replace it later, when costs are lower. The willingness of PHEV owners to allow this, and the benefits to them of doing so, need to be assessed. Conditions and terms under which this might be feasible and beneficial need to be developed. Alternatively, a charged PHEV might be used to provide electric power to a home during a blackout. It would be useful to know the viability of these options.

4

Scenario Analysis

The costs and consequences of deploying PHEVs into the U.S. market were estimated by analyzing two PHEV market penetration rates, the Maximum Practical scenario and the Probable scenario. The impacts on fuel consumption and well-to-wheel CO_2 emissions were then calculated using a modified version of the model developed for the 2008 Hydrogen Report (NRC, 2008). Because PHEVs will be substantially more expensive than HEVs, which are in turn more expensive than conventional vehicles, subsidies will be necessary to achieve these penetration rates, at least until vehicle costs decline sufficiently to be offset by the lower costs of driving on electricity. These subsidies are calculated for the two penetration scenarios using the expected vehicle costs from Chapter 2.

The Reference Case developed in the 2008 Hydrogen Report is used for comparing PHEVs in this report. Retaining that Reference Case for the present study allowed comparison with scenarios in the 2008 Hydrogen Report, although it precluded updating some of the numbers there such as those for oil prices, which were higher in the 2009 Annual Energy Outlook. Forecasts of energy supply and demand over such a long period are in any case highly uncertain. In particular, it is quite possible that the world production of conventional crude oil will reach a maximum during the intervening period and then go into decline, as forecast by a number of individuals and organizations.[1] Other analysts predict that supplies will be ample,[2] but if worldwide oil shortages cause dramatic oil price escalations during the period covered in this analysis, the world market for light-duty vehicles will change dramatically.

[1]For example, see U.K. Industry Taskforce on Peak Oil and Energy Security, 2008; J. Schlindler et al., 2008; R.A. Kerr, 2008; and Reuters, 2009.

[2]See the Energy Information Administration's *Annual Energy Outlook 2009*, available at http://www.eia.doe.gov/oiaf/aeo/index.html, or ExxonMobil's *The Outlook for Energy: A View to 2030*, available at http://www.exxonmobil.com/corporate/files/news_pub_2008_energyoutlook.pdf.

SCENARIO DESCRIPTIONS

In addition to the Reference Case, three other scenarios from the 2008 Hydrogen Report—hydrogen success (Case 1), advanced efficiency of conventional HEVs and nonhybrid vehicles (Case 2), and biofuels (Case 3)—are compared with the two PHEV scenarios. Portfolio cases that combine PHEVs with advanced efficiency and biofuels are also analyzed.

All scenarios describe possible futures for the U.S. light-duty vehicle fleet out to 2050 with the same total number of vehicles and vehicle-miles traveled. However, the vehicle mix over time is different for each scenario as described below.

Cases from the 2008 Hydrogen Report

Reference Case

This case was based on projections out to 2030 in the high oil price scenario in the Annual Energy Outlook 2008 (EIA, 2008) for the number of vehicles and their fuel consumption, oil prices, and other factors. The committee extended the curves to 2050. As shown in Figure 4.1, conventional gasoline internal combustion engine vehicles (ICEVs) continue to dominate the light-duty sector. Gasoline HEVs gain about 10 percent fleet share by 2050. The fuel economy of these vehicles follows projections from the EIA Annual Energy Outlook 2008, meeting fuel economy standards that rise until 2020, with only modest improvements in fuel economy beyond this time. HEVs reach 44.5 mpg in 2050, while nonhybrids reach 31.7 mpg, as shown in Figure 4.2.

Hydrogen

Hydrogen fuel cell vehicles (HFCVs) are introduced beginning in 2012, reaching 10 million on the road by 2025 and 60 percent of the fleet by 2050. Initially, hydrogen is

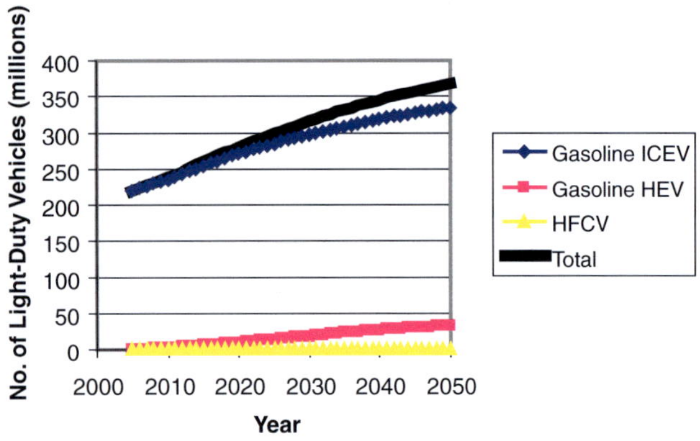

FIGURE 4.1 Number of light-duty vehicles in the fleet for the Reference Case. SOURCE: NRC, 2008.

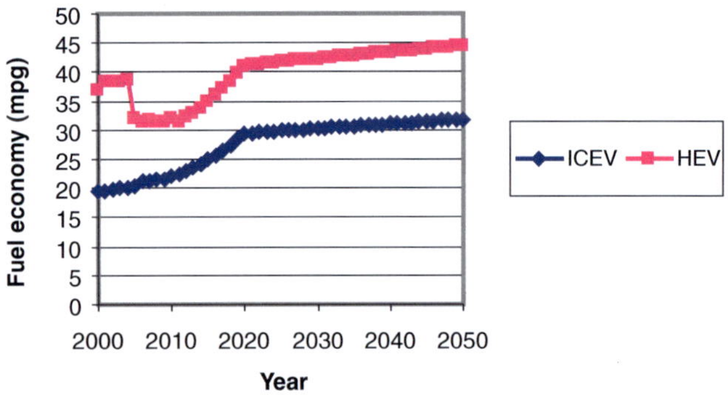

FIGURE 4.2 On-road fuel economy for vehicles in the Reference Case. SOURCE: NRC, 2008.

produced from natural gas, but over time energy sources that emit less carbon are used to produce hydrogen (biomass gasification and coal gasification with carbon capture and sequestration).

Efficiency

Improvements in engines and other vehicle technologies continue to be implemented past 2020. The fuel economy of ICEVs and HEVs is assumed to increase according to the following schedule:

- 2.7 percent per year from 2010 to 2025,
- 1.5 percent per year from 2026 to 2035, and
- 0.5 percent per year from 2036 to 2050.

In addition, HEVs become much more important, comprising 60 percent of the fleet by 2050. The fleet mix is shown in Figure 4.3. Fuel economy for both types of vehicles approximately doubles by 2050 (Figure 4.4), when HEVs average 60 mpg and ICEVs are at 42 mpg.

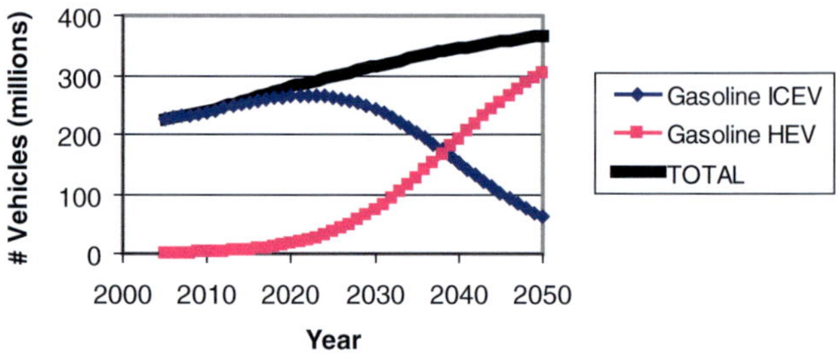

FIGURE 4.3 Types and numbers of light-duty vehicles for the Efficiency Case. SOURCE: NRC, 2008.

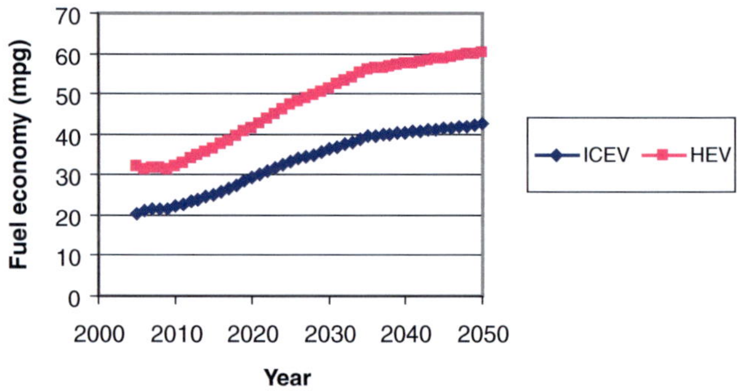

FIGURE 4.4 Fuel economy of new light-duty vehicles for the Efficiency Case. SOURCE: NRC, 2008.

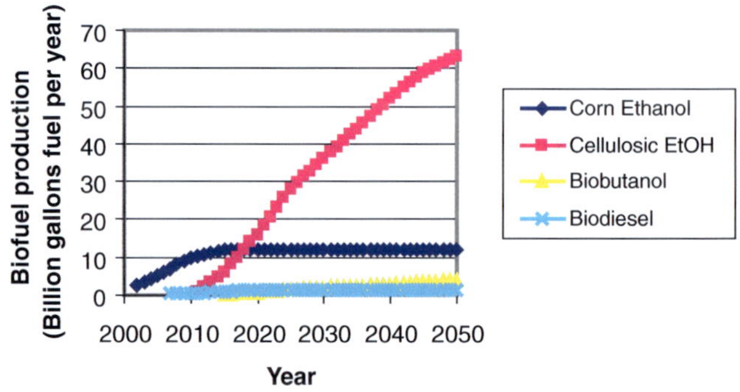

FIGURE 4.5 Biofuel supply for the Biofuels-Intensive Case. SOURCE: NRC, 2008.

Biofuels

Biofuels are introduced at a rapid rate, reaching 75 billion gallons per year in 2050 (Figure 4.5). Production of corn ethanol levels off, but cellulosic ethanol grows rapidly, reducing carbon emissions (well-to-wheels greenhouse gas [GHG] emissions for cellulosic ethanol are only 15 percent those of gasoline). Competition with food crops and indirect land use impacts on GHG emissions are not considered in this analysis.

BOX 4.1
Manufacturers' Announced Plans for Electric Vehicles (Partial List)

BMW	PHEV 50-km (31-mile) range in electric mode. 98 lithium polymer cells with a 2.5-hour charge time.
Ford	PHEV scheduled for 2011.
General Motors/Chevrolet	Volt scheduled for release in late 2010. PHEV 40-mile range in electric mode. 8-hour charge time at 120 V (3 hr at 240 V). 220 Li-ion battery cells.
Honda	PHEV scheduled for 2015.
Toyota	PHEV scheduled for 2012.
Nissan	EV scheduled for 2011.
Mitsubishi	EV released in Japan in 2009.
Hyundai	PHEV 40-mile range in electric mode.
BYD Co. (Chinese)	PHEV 60-mile range in electric mode. Special charging stations will charge to 70 percent in 10 minutes.

PHEV Cases

In these two scenarios, PHEVs replace some of the vehicles in the Reference Case which is otherwise unchanged.

Maximum Practical Penetration

The Maximum Practical scenario uses the same annual sales rate for PHEVs as the Hydrogen Case for HFCVs except that sales are initiated in 2010, 2 years earlier.[3] Auto companies are currently scheduling both PHEV-10 and PHEV-40 vehicles for introduction in that year (see Box 4.1).

This scenario assumes that manufacturers are able to rapidly increase production and that consumers find these vehicles acceptable. The Maximum Practical scenario would lead to approximately 240 million PHEVs on the road by 2050, the end of the scenario period, as shown in Figure 4.6. Such rapid penetration would require strong policy intervention because PHEVs will cost significantly more than comparable ICEVs and HEVs. At current gasoline prices, the fuel savings will not offset the higher initial cost. This policy intervention could be made in a variety of ways: mandates to vehicle manufacturers; subsidies to the purchasers

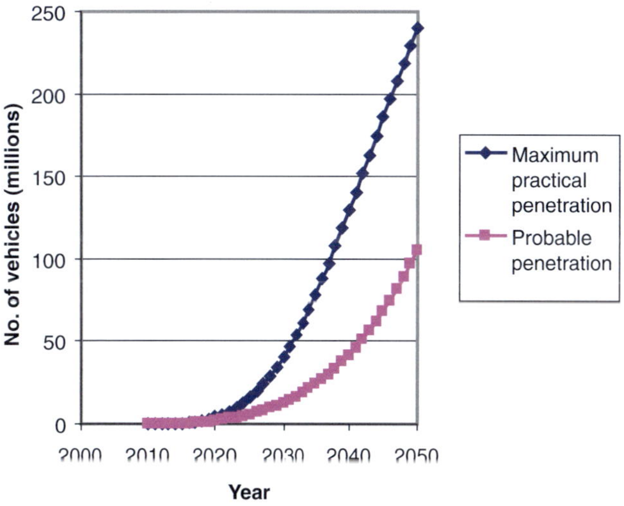

FIGURE 4.6 Penetration of PHEVs in the U.S. light-duty fleet.

[3]The PHEV scenarios are described in more detail in Appendix C.

BOX 4.2
Factors Affecting Deployment and Impact

PHEVs will not significantly reduce oil consumption and carbon emissions until there are tens of millions of them on the nation's roads. Whether and when this might happen is highly uncertain, in part because the following factors are still uncertain at this time:

- The rate at which the cost of batteries can be reduced,
- Future cost and fuel economy of HEVs and advanced conventional vehicles,
- Future costs and potential disruptions to the supply of oil,
- Changes in government policies, in particular fuel economy standards, carbon restrictions, and subsidies for PHEVs,
- The availability of a suitable place to charge the batteries and the potential additional cost of installing a new electric circuit,
- Consumer acceptance of the additional cost of PHEVs relative to competing vehicles of comparable size and performance, especially HEVs, and their willingness to accept vehicles that must be plugged in virtually every day,
- Resale value—some current HEVs have shown low depreciation rates, but if consumers are concerned that batteries will not last the life of the car, or that later owners, who are more likely to live in apartments, will not have access to a place to charge the vehicle, then they will be less likely to buy PHEVs, and
- Large vehicle fleets may be appropriate for PHEVs if the costs are reasonable. Many such fleets, among them the massive federal fleet, are largely used locally for short distances, with the vehicles returning to a central location at night, prime conditions for PHEVs. In 2008 the federal fleet numbered about 645,000 vehicles, led by the Department of Defense (30 percent) and the U.S. Postal Service (34 percent). In fiscal 2008, federal agencies ordered over 70,000 vehicles, approximately 11 percent of the total federal fleet. About 80 percent of these were light-duty trucks and passenger vehicles.[1]

The impact PHEVs will have for any specific growth rate also is uncertain:

- How many miles per year will actually be driven on battery power, given that many people do not drive significant distances every day and, even when they do drive, may not charge their vehicles every day, and
- Carbon emissions per kilowatt-hour used varies widely across the country and with the time of day when it is generated, and projections for the future are even more varied.

Resolving such uncertainties was not possible in this study, but it will be important to consider them when planning for the future of PHEVs.

[1]DOE Office of Energy Efficiency and Renewable Energy, ***Transportation Energy Data Book,*** Edition 28. Available at http://cta.ornl.gov/data/tedb28/Edition28_Chapter07.pdf.

of PHEV (perhaps greater than the current federal tax credit of $7,500) to offset the additional costs of the vehicles; and taxes or restrictions on fuel, but these are beyond the scope of this study.[4]

This scenario uses the optimistic technology costs discussed in Chapter 2. If costs fail to decline to those levels, this scenario would be prohibitively expensive.

Probable Penetration

The Probable scenario represents a PHEV market penetration that the committee judges to be more likely in the absence of strong market-forcing policies to supplement the policies already in place. It also starts in 2010, but market penetration is slower than in the Maximum Practical scenario, reflecting factors described in Box 4.2. PHEVs rise to 3 percent of new light-duty vehicles entering the U.S. vehicle fleet by 2020 and to 15 percent by 2035.[5] This pace would lead to 110 million PHEVs on the road by 2050, as shown in Figure 4.6.

The Probable scenario assumes the continuance of current policy incentives, which are inadequate to achieve the penetration rate in the Maximum Practical scenario. Vehicles

[4]Alternatively, a sharp and prolonged rise in the price of petroleum could have the same motivational effect. However, the adverse consequences of such an event for the health of the economy could leave consumers without sufficient financial resources to purchase large numbers of PHEVs.

[5]The committee based its estimate on estimates in the America's Energy Future (AEF) Committee report, which drew on "historical case studies of comparable technology changes" (NAS-NAE-NRC, 2009, p. 165). The AEF study estimated that PHEVs would represent 1 to 3 percent in 2020 and 7 to 15 percent in 2035.

are more expensive in the Probable scenario because it uses the probable technology costs discussed in Chapter 2.

PHEV Portfolio Cases

Two additional cases combine PHEVs with other technologies to investigate how they may work together.

PHEV + Efficiency

This scenario is the same as the PHEV Maximum Practical Case above, except that the fuel economy of the rest of the fleet (ICEVs and HEVs) improves as in the Efficiency Case. The vehicle mix is shown in Figure 4.7. PHEVs make up 65 percent of the fleet in 2050, but 19 percent are HEVs and only 16 percent are conventional nonhybrid vehicles. This case is actually more realistic than the Maximum Practical case, because it makes little sense to invest in expensive PHEVs unless the more cost-effective efficiency measures are implemented first.

PHEV + Efficiency + Biofuels

This case adds biofuels to the PHEV + Efficiency Case above, replacing some of the gasoline used by ICEVs, HEVs, and PHEVs. The vehicle mix is the same (Figure 4.7), but the vehicles use significant amounts of biofuels instead of gasoline.

PHEV Characteristics

The PHEV-10 and PHEV-40 are the only vehicles modeled in this report. These are both midsize cars,[6] as are the vehicles in the 2008 Hydrogen Report. Modeling a range of light-duty vehicles was beyond the resources of this study. Therefore, the results should be viewed as approximations. All-electric vehicles were not included in this study.

PHEVs are complicated to model because some of their energy comes from gasoline and some from the grid. The fraction of vehicle miles traveled on electricity rather than gasoline and the consumption of electricity and fuel over a drive cycle are influenced by several factors. The methodology used to calculate gasoline and electricity consumption is detailed in Appendix C. Energy consumption for all the vehicles discussed here is shown in Table 4.1.

PHEV costs are as discussed in Chapter 2. The retail prices that might be expected (40 percent greater than manufacturing costs[7]) are shown in Table 4.2. Figure 4.8 compares

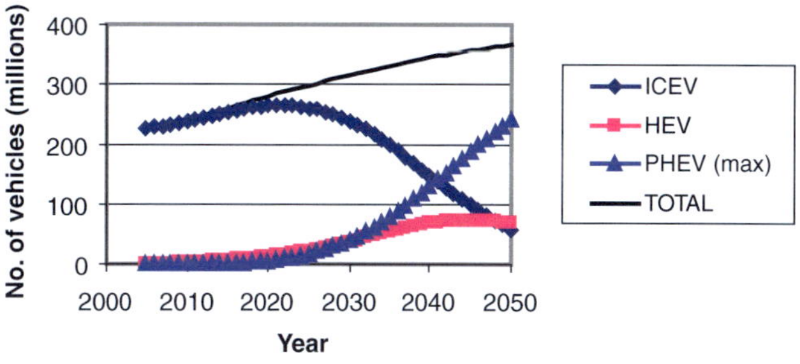

FIGURE 4.7 Number of vehicles for the Portfolio Cases, a mix of PHEVs and efficient ICEVs and HEVs, introduced at the Maximum Practical rate.

these prices to those of the Reference Case vehicle. PHEV costs are significantly higher throughout the time frame of this study (2010 to 2050).

The prices of gasoline and electricity are shown in Figure 4.9. Gasoline prices rise significantly, but electricity prices do not and are here treated as constant at 8 cents per kWh for simplicity, a rate slightly lower than the national average to reflect promotional or time-of-use rates at night. A case analyzing the effect of higher electricity prices is discussed in Appendix C.

TRANSITION COSTS

Investments will be required for PHEVs to reach cost competitiveness with the Reference Case gasoline vehicle. This transition cost analysis is similar to that in the 2008 Hydrogen Report: It examines the annual cash flows to find the total investment required. Cost competitiveness is achieved in the break-even year, when the total incremental costs for all the new PHEVs bought that year is balanced by the annual fuel savings for all PHEVs on the road in comparison to the reference vehicles.[8]

Investment costs in this case are basically government buydowns or subsidies to cover some or all of the incremental costs in order to encourage the public to buy the vehicles. Manufacturers may at first charge less than the price needed to cover their costs when only a few vehicles are being sold, but that is unlikely to be feasible after a few years at the penetration rates envisioned here. In addition, costs would

[6]The fuel economy and electric use of the modeled mid-sized PHEV cars are similar to results from Simpson (2006). The energy use is somewhat higher than projections for smaller electric vehicles such as the Volt.

[7]To make a profit, manufacturers must pass on the cost of the components they buy for their products and some fraction more. These additional costs are needed to cover their design, installation, and warranty costs, among other things.

[8]The cash flow analysis is not a discounted life-cycle cost analysis. It is an estimate of the subsidies required each year to make PHEVs appear cost-effective to the consumer and compares those to the fuel savings from all the PHEVs on the road that year. Note that PHEVs are compared to the reference vehicle, which is a nonhybrid. Consumers considering a PHEV are much more likely to compare it to an equivalent HEV, which will be significantly cheaper, get very good fuel economy, and not require daily plugging in. The committee decided to use the same Reference Case as in the 2008 Hydrogen Report to allow comparability with that study. If an HEV had been used as the reference vehicle, the incremental costs would have been lower, but so would have been the fuel savings, as shown in Figure 4.11.

TABLE 4.1 Energy Requirements of Midsized Vehicles

	PHEV-10	PHEV-40	HEV		Conventional Non-HEV	
Control strategy, charge-depleting mode	Blended	Battery only	—		—	
Gasoline consumption, gal/100 mi			Ref. Case.	Efficient	Ref. Case.	Efficient
2010	2.5	1.4	3.1	3.1	4.5	4.5
2020	1.9	1.1	2.4	2.4	3.4	3.4
2035	1.4	0.8	2.3	1.8	3.3	2.5
2050	1.3	0.7	2.2	1.7	3.1	2.4
Electricity consumption, Wh/mi						
2010	99	251	—		—	
2020	76	193				
2035	57	143				
2050	53	133				

NOTE: HEV and conventional non-HEV data from NRC (2008). PHEV numbers derived from Kromer and Heywood (2007). The electricity data are different from those discussed in Chapter 2 because these are more representative of a diverse fleet, and they decline over time as the vehicle becomes more efficient. Estimates of PHEV electricity consumption vary widely. Gasoline and electricity consumption for new cars, on-road, averaged over drive cycle. PHEV-10 gasoline consumption = 81 percent of efficient HEV. PHEV-40 gasoline consumption = 45 percent of efficient HEV.

TABLE 4.2 Estimated Retail Prices of PHEVs Incremental to Retail Price of Reference Case Gasoline Car (dollars)[a]

	PHEV-10[b]			PHEV-40[c]		
	Optimistic	Probable	DOE Goal[d]	Optimistic	Probable	DOE Goal[e]
2010	7,700	8,800		19,800	25,400	
2020	5,600	6,300	4,500[f]	13,500	17,000	7,600[g]
2030	5,100	5,700		12,300	15,500	
OEM battery cost, $ per usable kWh	720	950		720	1,000	

[a]Retail price = 1.4 × OEM cost (see Table 2.7). An efficient ICEV would cost, at retail, about $1,000 more than the Reference Case gasoline vehicle. The retail price for an efficient HEV would be about $2,000 more than the Reference Case gasoline vehicle, as discussed in the 2008 Hydrogen Report.

[b]Battery size (energy used) = 2.0 kWh (nameplate 4 kWh).

[c]Battery size (energy used) = 8.0 kWh (nameplate 16 kWh).

[d]Goal for OEM cost of battery ($500/usable kWh).

[e]Goal for OEM cost of battery ($300/usable kWh).

[f]Cost of PHEV-10 meeting DOE goals assumes that optimistic 2030 vehicle parameters are achieved, but the battery costs $500/kWh instead of $720/kWh. For a 2 kWh battery this subtracts ($720 – $500/kWh) × 2 kWh = $440 from the OEM cost of the vehicle. Accounting for a retail price mark-up factor of 1.4, the added cost is about 1.4 × $440, or ~$600. So the retail price of the vehicle meeting the DOE battery goal is $5,100 (optimistic 2030 case) – $600 (cost reduction for lower cost battery) = $4,500.

[g]Cost of PHEV-40 meeting DOE goals assumes that optimistic 2030 vehicle parameters are achieved, but the battery costs $300/kWh instead of $720/kWh. For an 8 kWh battery this subtracts ($720 – $300/kWh) × 8 kWh = $3,360 from the OEM cost of the vehicle. Accounting for a retail price mark-up factor of 1.4, the added cost is about 1.4 × $3,360 ~ $4,700. So the retail price of the vehicle meeting the DOE battery goal is $12,300 (optimistic 2030 case) – $4,700 (cost reduction for lower cost battery) = $7,600.

be incurred for deploying charging facilities for PHEVs.[9] Unlike the analysis in the 2008 Hydrogen Report, investment costs here do not include research and development or any energy supplier costs, even though these are nonzero for PHEVs.

The cash flow analysis is described in Appendix C. Table 4.3 summarizes the results for the two PHEVs under the Maximum Practical penetration scenario. It also shows the results for a 30/70 mix of PHEV-40s and PHEV-10s, showing the effect of two different kinds of PHEVs in the market.

These results depend to a significant extent on the assumptions that go into the analyses. To explore these, Appendix C also includes a sensitivity analysis for the PHEV price increment, oil price, and electricity price. Break-even

[9]Capital costs for in-home charging facilities are not explicitly added to the electricity cost or vehicle price in the cash flow analysis. These would likely would have a very small impact on the breakeven year or buydown cost (see Appendix C for sensitivity analysis).

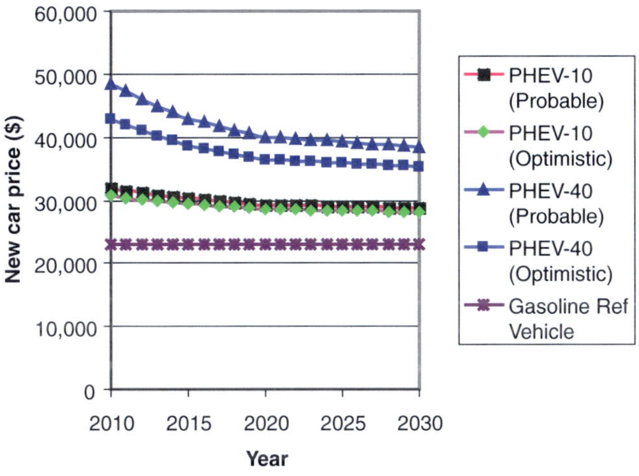

FIGURE 4.8 Retail prices for PHEVs for probable and optimistic rates of technology progress, compared to the Reference Case vehicle (conventional ICEV).

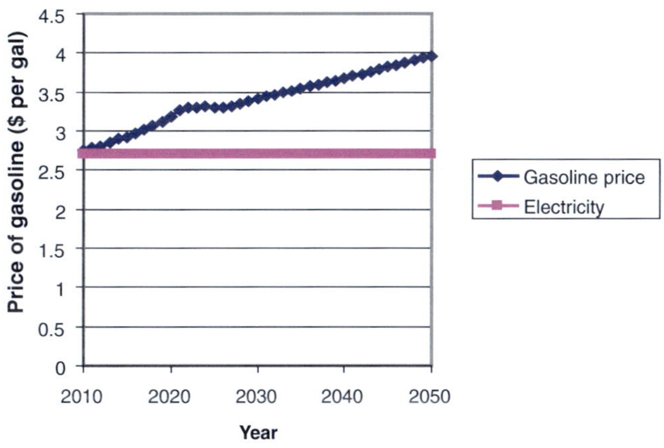

FIGURE 4.9 Price of gasoline over time and electricity price of 8 cents per kilowatt-hour. SOURCE: EIA, 2008 (gasoline price, high).

year and buydown costs are very sensitive to oil price and PHEV incremental price, as shown in Table 4.3, but much less so to electricity price. If gasoline costs twice as much as shown in Figure 4.9, with optimistic technology progress the PHEV-40 reaches breakeven in 2020 instead of 2028 and the PHEV-40 in 2025 instead of 2040. Similarly, if the even more optimistic DOE goals for battery costs are met by 2020, breakeven for the PHEV-10 is in 2020 and the PHEV-40 in 2024 (with Reference Case oil prices). These results underscore the need for battery technology breakthroughs.

Figure 4.10 illustrates the various cash flows for the PHEV-10 at the Maximum Practical penetration rate as follows:

- The vehicle cost difference is the difference between the pricc of a conventional gasoline vehicle and a PHEV (see Figure 4.8), summed over all the new PHEVs sold that year. This is negative because PHEVs always cost more than conventional vehicles. It is small at first even though the cost differential is large because only a few PHEVs are sold. It continues to grow as more vehicles are sold each year.
- The fuel cost difference is the annual difference in fuel costs for all PHEVs currently in the fleet and the same number of comparable conventional vehicles. Electricity is generally less costly than gasoline on a cents per mile basis (Figure 4.9), so this difference is positive.
- Cash flow combines these two curves to represent the economy-wide cost per year of pursuing a PHEV introduction plan. It starts out negative because all the PHEVs sold in a year are much more expensive that the conventional vehicles they replace, but there are few PHEVs in the fleet producing fuel savings. Cash flow goes positive in 2028 (the break-even year) because the total fuel savings exceed the purchase cost differential of the PHEVs sold that year.
- Cumulative cash flow is a year-by-year summation of the annual cash flow over time (starting in 2010). It provides a tally of the total funds that would have to be invested to make PHEVs competitive. At first, there is a negative cash flow (early PHEVs cost more than gasoline cars), but, as PHEV-10 costs come down, the negative cash flow bottoms out in 2028 at a minimum of about $33 billion, when about 24 million PHEV-10s have been produced. This minimum is the buydown investment that must be supplied to bring the PHEV-10 to cost competitiveness.

Most of the negative cash flow is due to the high price of the first few million PHEVs. This is not surprising since PHEVs initially cost a lot more than conventional vehicles. The subsidy that might be needed by automakers or buyers is the sum of the difference in costs between PHEV-10s and conventional cars, each year between vehicle introduction in 2010 and breakeven in 2028. This cumulative difference in vehicle first cost for PHEVs (as compared to a reference vehicle) is about $133 billion (averaged over the 2010-2028 buydown period, this is about $5,400 per car, or an average of $7.4 billion per year for 18 years).

Table 4.3 shows that PHEV-40s have a significantly higher transition cost than PHEV-10s because the larger battery is very expensive. The mixed cases lie between the PHEV-10 and PHEV-40 cases. Although the 30 percent of the PHEV fleet made up of PHEV-40s is costly, this is offset by the lower cost of the more numerous PHEV-10s. The break-even time is about 5 years earlier, and the buydown cost is less than for a pure PHEV-40 case.

Table 4.4 compares the transition costs for the 30/70 mix of PHEVs with the two penetration cases. Interestingly, the slower market penetration of the Probable Case gives a lower overall transition cost than the Maximum Practical Case. In the maximum practical cases, more PHEVs are bought

TABLE 4.3 PHEV Transition Times and Costs

	PHEV-40	PHEV-40	PHEV-40 High Oil[a]	PHEV-10	30/70% PHEV-40/10 Mix	
Penetration Rate:	Maximum Practical	Maximum Practical	Maximum Practical	Maximum Practical	Maximum Practical	Probable
Technical Progress:	Optimistic	DOE Goal[b]	Optimistic	Optimistic	Optimistic	Probable
Break-even year[c] (annual cash flow = 0)	2040	2024	2025	2028	2032	2034
Cumulative subsidy to break-even year (billion $)[d]	408	24	41	33	94	47
Cumulative vehicle retail price difference until the break-even year (billion $)[e]	1,639	82	174	51	363	—
Number of PHEVs sold to break-even year (millions)	132	10	13	24	48	20

[a]Assumes oil costs twice that in the base case, or $160/bbl in 2020, giving results similar to meeting DOE's cost goals.

[b]Assumes DOE technology cost goal ($300/kWh) for the PHEV-40 is met by 2020, showing the importance of technology breakthroughs as discussed in Chapter 2 and Appendix F. Reducing costs this rapidly would significantly reduce subsidies and advance the break-even year relative to the Optimistic Technical Progress cases.

[c]Year when annual buydown subsidies equal fuel cost savings for fleet.

[d]Does not include infrastructure costs for home rewiring, distribution system upgrades, and public charging stations which might average over $1000 per vehicle.

[e]Cost of PHEVs minus the cost of Reference Case cars.

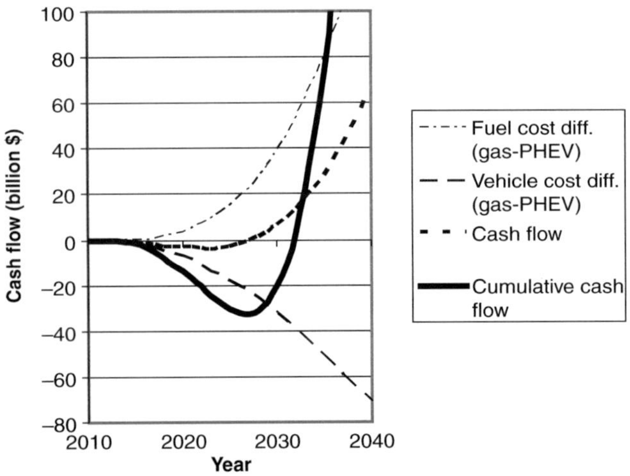

FIGURE 4.10 Cash flow analysis for PHEV-10, Maximum Practical Case, Optimistic technical assumptions. The break-even year is 2028, and the buydown cost is $33 billion.

earlier, when they are more expensive, leading to higher transition costs.[10] Table 4.4 also compares the transition costs of fuel cell vehicles as estimated in the 2008 Hydrogen Report.

OIL CONSUMPTION

Fuel consumption was calculated for the two penetration cases with the model used in the 2008 Hydrogen Report, modified to account for PHEV characteristics and the use of two different fuels. Results are compared with the Reference Case in Figure 4.11. For the Maximum Practical Case, the PHEV-40 cuts gasoline use by 55 percent by 2050, and PHEV-10s cuts it 40 percent.[11]

However, much of the savings achieved with PHEVs could also be attained by HEVs, as shown in Figure 4.12, the

[10]The committee used the same rate of cost reductions (Table 2.6) over time for both penetration rates. As discussed in Chapter 2, most cost reductions are likely to be from technology improvements. While economies of scale will be realized, these are likely to be modest (because Li-ion battery production is already very high) and may be offset by increases in the cost of materials with greater demand.

[11]The terms "oil" and "gasoline" are used interchangeably in this report. While not strictly accurate, reducing consumption of gasoline by one gallon will reduce demand (and imports) of oil by close to one gallon once adjustments at the refinery are accounted for.

TABLE 4.4 Comparison of Transition Costs for PHEV and HFCV Cases

	30/70 PHEV-40/PHEV-10 Mix		HFCV	
Penetration Rate	Maximum Practical	Probable	H_2 Success	H_2 Partial Success
Break-even year[a]	2032	2034	2023	2033
Cumulative cash flow difference (PHEV-gasoline reference car) to break-even year[b]	$94 billion	$47 billion	$22 billion	$46 billion
Cumulative vehicle retail price difference (AFV-gasoline reference car) to break-even year	$363 billion	$179 billion	$40 billion	$92 billion
Number of PHEVs sold to break-even year (millions)	48	20	5.6	10.3
Infrastructure cost	$48 billion	$20 billion	$8 billion	$19 billion
	(In-home charger $1,000 per car)	(In-home charger $1,000 per car)	(H_2 stations for first 5.6 million HFCVs)	(H_2 stations for first 10.3 million HFCVs)

[a]Year when annual buydown subsidies equal fuel cost savings for fleet.

[b]Does not include infrastructure costs for home rewiring, distribution system upgrades, and public charging stations which might average over $1000 per vehicle.

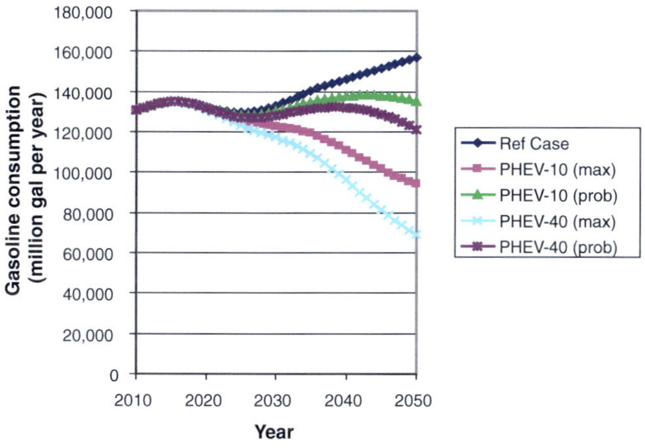

FIGURE 4.11 Gasoline consumption for PHEV-10s and PHEV-40s introduced at Maximum Practical and Probable penetration rates shown in Figure 4.6.

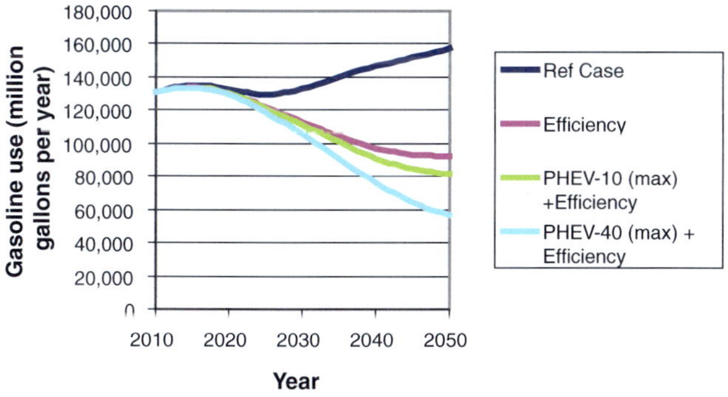

FIGURE 4.12 Gasoline use for the Reference Case and the Efficiency Case and when PHEVs are included in an already highly efficient fleet, as shown in Figure 4.7.

first Portfolio Case.[12] Figure 4.12 also shows the Efficiency Case from the 2008 Hydrogen Report. Gasoline use is cut by about 40 percent, mainly with advanced HEVs and no PHEVs. When PHEVs are introduced into this fleet instead of the Reference Case fleet, PHEV-10s reduce fuel consumption by an additional 7 percent and PHEV-40s by 20 percent beyond the Efficiency Case as shown by the two lower curves in Figure 4.12 for the Maximum Practical Case.

The impact on oil consumption of adding biofuels to this fleet is shown in Figure 4.13, the final Portfolio Case. Combining biofuels with advanced efficiency, including HEVs, can cut oil consumption by about 65 percent compared with the Reference Case by 2050, as shown in Figure 4.13. Adding PHEV-10s to that mix can reduce consumption by another 7 percent, while PHEV-40s could account for 23 percent. Figure 4.13 also shows the results from the 2008 Hydrogen

[12]A 40-mpg HEV would use 375 gallons in 15,000 miles. As noted in Table 4.1, the equivalent PHEV-10 would use 81 percent as much fuel, or 304 gallons for the same distance, a savings of just 71 gallons. The PHEV-40, which uses just 45 percent of the fuel of the equivalent HEV will do better, saving 206 gallons. The most gasoline a PHEV-10 can save relative to a 40-mpg HEV is one quart per charge (the 10 miles driven on electricity would require that much more gasoline in the HEV). If it is driven at least 10 miles and then recharged every day, the PHEV-10 would save a total of 91 gallons per year, but many drivers will not adhere to such a regular schedule. Charging more than once a day could increase these savings, but that would probably apply to relatively few vehicles, especially in the early years, when public charging stations are rare. Results from the North American PHEV Demonstration project, involving over 100 Toyota Prius conversions to PHEVs (approximately equivalent to the PHEV-10), measured an average fuel economy of 50 mpg. With the battery pack depleted or turned off, mileage was 44 mpg (DOE/EERE, 2009), about what a conventional Prius would achieve. While a converted Prius might not fully reflect the performance of optimized PHEV, these tests show that in ordinary driving, a PHEV-10 is unlikely to provide large fuel savings. Furthermore, HEVs are expected to increase their mileage, perhaps to an average of 60 mpg by 2050, reducing the benefits of PHEVs.

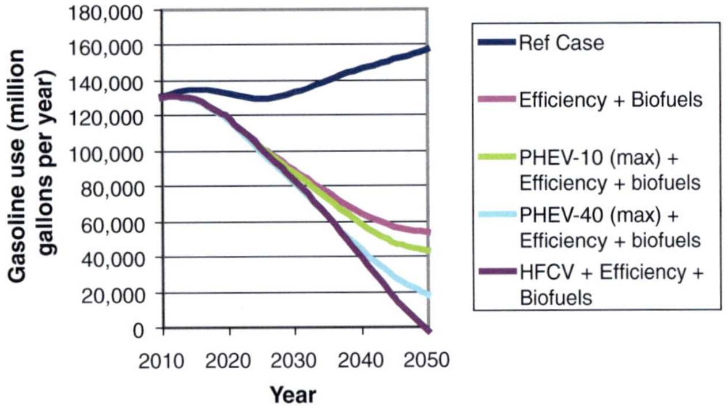

FIGURE 4.13 Gasoline use for scenarios that combine efficiency, biofuels, and either PHEVs or HFCVs.

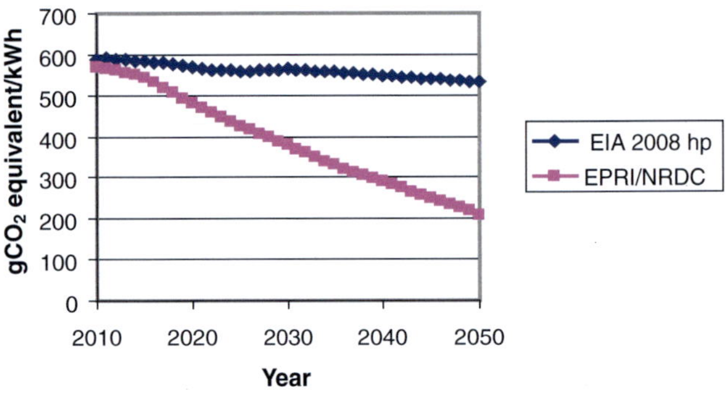

FIGURE 4.14 GHG emissions from the future electric grid. SOURCES: EPRI/NRDC estimates from EPRI/NRDC (2007), and EIA estimates from Annual Energy Outlook, 2009 (EIA, 2009a).

Report when HFCVs are combined with efficiency and biofuels, which could completely eliminate gasoline consumption by the light-duty vehicle fleet by 2050.[13]

CARBON DIOXIDE EMISSIONS

PHEVs emit less CO_2 because they use less gasoline than conventional vehicles, but generating the electricity that replaces the gasoline usually results in emissions. Thus, total GHG emissions from PHEVs depend on the composition of the electric grid and on the time of day for charging.[14]

The committee analyzed two projections for the grid:

- A business-as-usual case, starting with the high price case from the Annual Energy Outlook (EIA, 2008) and extended to 2050 using the same growth rate for electric sector CO_2 emissions;
- A low-carbon grid projection from a joint EPRI/NRDC study (EPRI/NRDC, 2007).

The carbon emissions per kilowatt hour for both grid scenarios are shown in Figure 4.14. These projections are discussed in more detail in Appendix C.

Figures 4.15 and 4.16 show CO_2 emissions under the two sets of grid conditions. Emissions under the EPRI/NRDC mix are significantly lower.[15] Figure 4.17 compares HFCVs to PHEVs, for the Maximum Practical Case with the low-carbon grid. HFCVs give a lower rate of GHG emissions than PHEV-10s, which still use a significant amount of gasoline. FCVs have lower emissions than PHEV-40s beyond about 2040. Low carbon emissions for both PHEVs and HFCVs

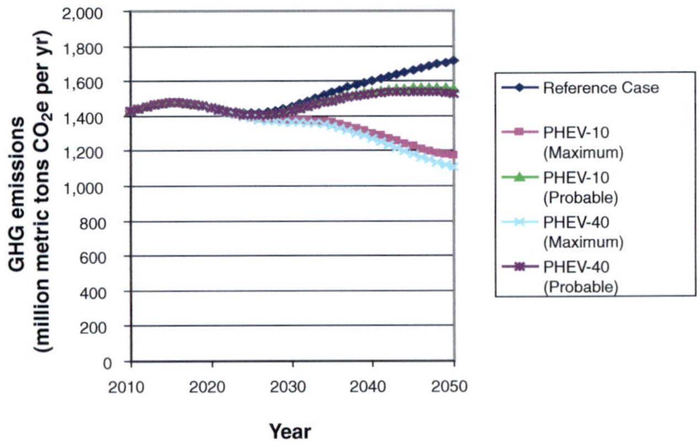

FIGURE 4.15 GHG emissions for PHEVs at the market penetrations shown in Figure 4.6 for the grid mix estimated by EIA. SOURCE: EIA, 2009a.

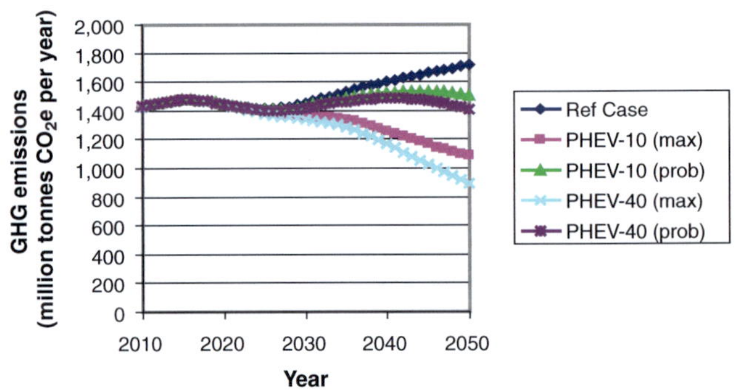

FIGURE 4.16 GHG emissions for PHEVs at the market penetrations shown in Figure 4.6 for the grid mix estimated by EPRI/NRDC. SOURCE: EPRI/NRDC, 2007.

[13]Some vehicles might still require gasoline or diesel fuel, but the use of biofuels to replace other uses of oil could more than compensate for this.

[14]This analysis did not include the additional GHG emissions from manufacturing a PHEV relative to a conventional vehicle.

[15]The reductions are only for the electricity used in the transportation sector. Total reductions from the electricity sector would be much greater than the difference between Figures 4.7 and 4.8.

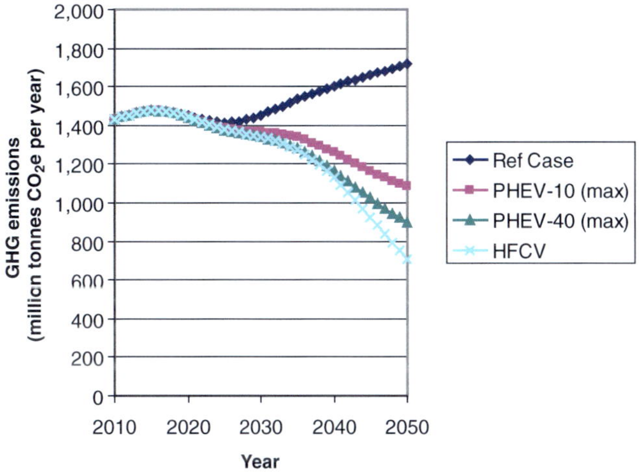

FIGURE 4.17 GHG emissions for cases combining ICEV Efficiency Case and PHEV or HFCV vehicles at the Maximum Practical penetration rate with the EPRI/NRDC grid mix.

depend on using lower carbon primary sources for electricity and hydrogen (see Appendix C).

For the first Portfolio Case, Figures 4.18 and 4.19 combine PHEVs at the Maximum Practical penetration rate with the Efficiency Case for the two grid mixes. For the EIA grid mix, there is very little difference in GHG emissions between the Efficiency Case, where no PHEVs are introduced, and the PHEV-10 and PHEV-40 cases. The benefit of PHEVs appears only when a lower carbon grid (the EPRI/NRDC grid mix) is used. This highlights the importance of low-carbon electricity for gaining the potential benefits of PHEVs. The HFCV case has significantly lower GHG emissions than either of the PHEV cases for a similar level of energy supply decarbonization. That is, well-to-tank carbon emissions for supplying hydrogen can be reduced by about two-thirds by 2050 (as in the 2008 Hydrogen Report), resulting in greater CO_2 reduction than when the electricity carbon emissions (g CO_2/kWh) are reduced by two-thirds by 2050 (as in the EPRI/NRDC grid case). This is true because HFCVs are somewhat more efficient than PHEVs on an energy per mile basis.[16]

Finally, the committee estimated GHG emissions for cases that combine efficiency, biofuels, and PHEVs or HFCVs for the two grid mixes (Figures 4.20 and 4.21). Again, the importance of a low-carbon grid is apparent for the PHEVs; the GHG emissions reduction in 2050 is about 55 percent for efficiency + biofuels, 59 percent (71 percent) for efficiency + biofuels + PHEV-10s (PHEV-40s), and 80 percent for efficiency + biofuels + HFCVs. With the

[16]Furthermore, the facilities to generate hydrogen from coal or natural gas will be new and use a process that can be adapted relatively easily to carbon capture. Retrofitting an existing pulverized coal electric plant (about 50 percent of current U.S. generating capacity) with carbon capture will be very expensive.

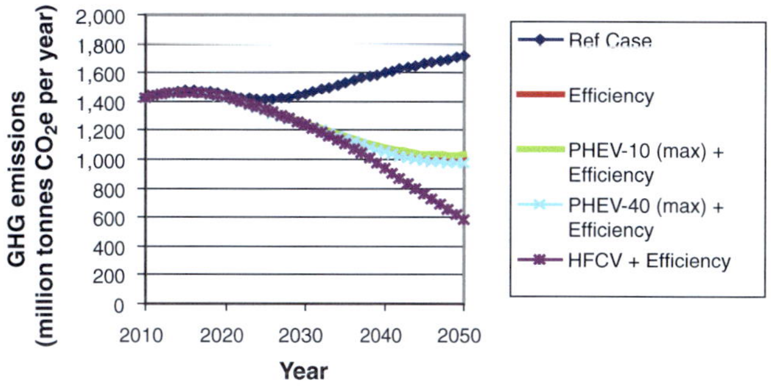

FIGURE 4.18 GHG emissions for cases combining ICEV Efficiency Case and PHEV or HFCV vehicles at the Maximum Practical penetration rate with the EIA grid mix.

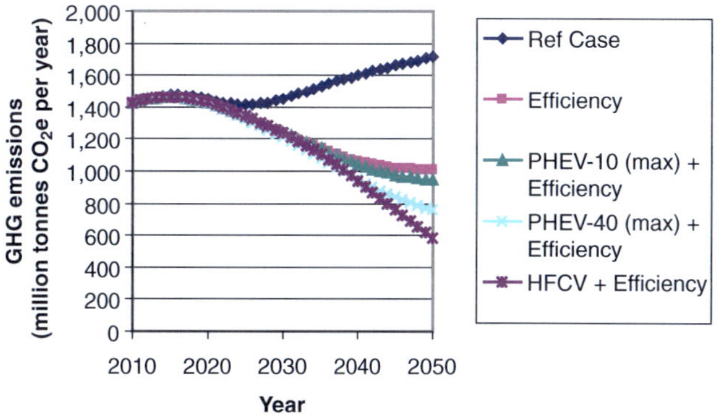

FIGURE 4.19 GHG emissions for cases combining the ICEV Efficiency Case and PHEV or HFCV vehicles for the EPRI/NRDC grid mix.

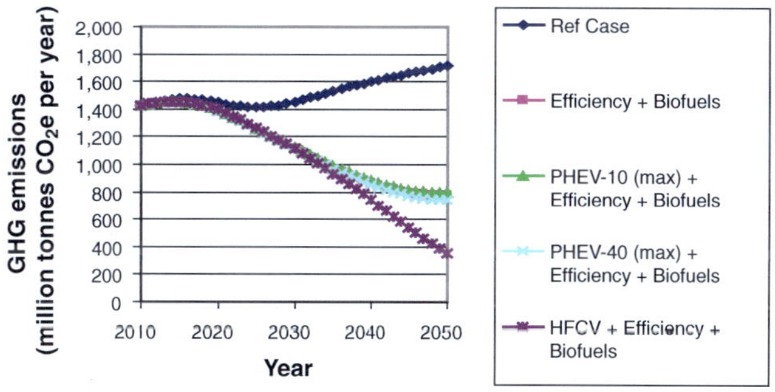

FIGURE 4.20 GHG emissions for scenarios combining ICEV Efficiency Case, Biofuels Case, and PHEVs or HFCVs for the EIA grid mix.

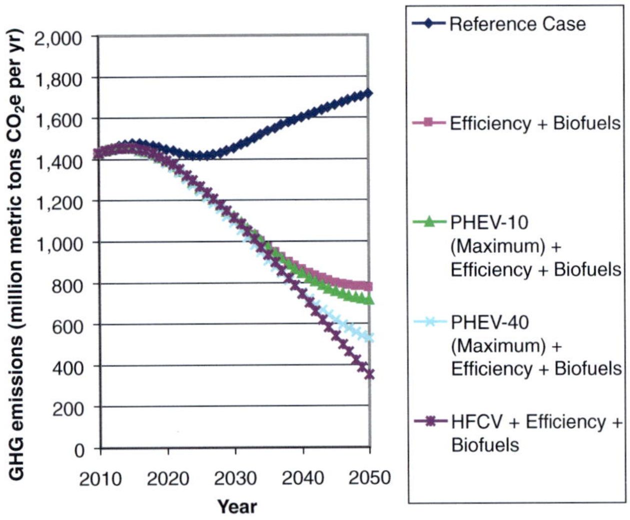

FIGURE 4.21 GHG emissions for scenarios combining ICEV Efficiency Case, Biofuels Case, and PHEVs or HFCVs for the EPRI/NRDC grid mix.

higher-carbon EIA grid, the GHG reduction with PHEV-10s (PHEV-40s) is about 55 percent (59 percent), about the same as for efficiency + biofuels.

SCENARIO SUMMARY

Societal Benefits of PHEVs

- GHG and oil reductions for PHEVs are small before 2025 because of the time needed for vehicles to penetrate the market.
- PHEV GHG benefits depend on the grid mix:
  —PHEV benefits are small compared with HEVs for the EIA grid.
  —With a low-carbon grid (EPRI/NRDC mix), introduction of PHEV-40s could significantly lower GHG emissions relative to HEVs.
- Increasing conventional vehicle efficiency alone (without PHEVs) can reduce oil use by about 40 percent in 2050 compared with the Reference Case. Adding PHEV-10s at the Maximum Practical rate can reduce oil use an additional 7 percent, while PHEV-40s can reduce it an additional 23 percent.
- Implementing efficiency plus biofuels reduces gasoline use by about 65 percent compared with the Reference Case. Adding PHEV-10s at the Maximum Practical rate can reduce oil use an additional 7 percent, while PHEV-40s can reduce it 23 percent.
- A portfolio approach incorporating efficiency, more use of HEVs and biofuels, as well as PHEVs, yields greater reductions in oil use and GHG.
- Long-term GHG and oil-use reductions are greater with HFCVs than PHEVs for similar levels of energy supply decarbonization (NRC Hydrogen scenario; EPRI/NRDC grid). If PHEVs are charged from the EIA grid, GHG emission reductions with PHEVs will be much less than with HFCVs.

Transition Costs

- Transition costs and timing to breakeven are similar for HFCVs and PHEV-10s, i.e., tens of billions of dollars total, spent over a 10-20 year period. This is less than the current corn ethanol subsidy of about $10 billion per year.
- Majority of transition cost (more than 80 percent) is for vehicle buydown. Average price subsidy needed for HFCVs and PHEV-10s over a 10-15 year transition period is similar, about $5000 to $6000 per car for PHEV-10s, and $7,000 to $9,000 per car for HFCVs.
- Transition costs for PHEV-40s are significantly higher than for PHEV-10s, because of higher vehicle first cost. Break-even year for the PHEV-40 is 2040 in the Optimistic Technology Case, but not until 2047 for the Probable Case, unless the oil price is high or the cost of batteries can be reduced rapidly.
- Slower Probable Case transition strategies sometimes have a lower overall transition cost than the Maximum Practical Case. This is true because the Maximum Practical Case buys large numbers of expensive early PHEVs.
- Transition costs are sensitive to oil prices and to vehicle cost increment, which depends on battery cost assumptions, but are not very sensitive to electricity price.
- Infrastructure costs for PHEVs might average $1000 per car for residential charging.
- Total infrastructure capital costs to breakeven are the same order of magnitude for PHEV-10s and HFCVs, although early infrastructure logistics are less complex with PHEVs.

Overall Messages from Scenarios

- Bringing PHEVs to cost-competitiveness will take several decades and require many billions of dollars in support. Transition costs for PHEV-40s are significantly larger than for PHEV-10s, but the reduction in gasoline consumption is greater also.
- GHG benefits of PHEVs depend on the grid mix. With a business-as-usual EIA grid mix, the benefits of PHEVs are similar to those for efficient gasoline HEVs. With a substantially decarbonized grid, PHEVs can save 4-16 percent more GHG emissions than efficient HEVs.
- The PHEV transition cost and timing results are sensitive to the oil price and the battery cost. But even with relatively high oil prices (AEO high oil price case $80-$120 per barrel) and achievement of aggressive battery goals (similar to the DOE goals), it will take 15-20 years and tens to hundreds of billions of dollars to bring PHEV-40s to commercial success.

5

Results and Conclusions

1. Lithium-ion battery technology has been developing rapidly, especially at the cell level, but costs are still high, and the potential for dramatic reductions appears limited. Assembled battery packs currently cost about $1,250 to $1,700 per kWh of usable energy ($625 to $850/kWh of nameplate energy). A PHEV-10 will require about 2.0 kWh and a PHEV-40 about 8 kWh even after the batteries have undergone expected degradation over time. Costs are expected to decline by about 35 percent by 2020 but more slowly thereafter. Projections of future battery pack costs are uncertain, as they depend on the rate of improvements in battery technology and manufacturing techniques, potential breakthroughs in new technology, possible relaxation of battery protection parameters as experience is gained, and the level of production, among other factors. Further research is needed to reduce costs and achieve breakthroughs in battery technology.

2. Costs to a vehicle manufacturer for a PHEV-40 built in 2010 are likely to be about $14,000 to $18,000 more than an equivalent conventional vehicle, including a $10,000 to $14,000 battery pack. The incremental cost of a PHEV-10 would be about $5,500 to $6,300, including a $2,500 to $3,300 battery pack. In addition, some homes will require electrical system upgrades, which might cost more than $1,000. In comparison, the incremental cost of an HEV might be $3,000.

3. PHEV-40s are unlikely to achieve cost-effectiveness before 2040 at gasoline prices below $4.00 per gallon, but PHEV-10s may get there before 2030. PHEVs will recoup some of their incremental cost, because a mile driven on electricity will be cheaper than a mile on gasoline, but it is likely to be several decades before lifetime fuel savings start to balance the higher first cost of the vehicles. Subsidies of tens to hundreds of billions of dollars will be needed for the transition to cost-effectiveness. Higher oil prices or rapid reductions in battery costs could reduce the time and subsidies required to attain cost-effectiveness.

4. At the Maximum Practical rate, as many as 40 million PHEVs could be on the road by 2030, but various factors (e.g., high costs of batteries, modest gasoline savings, limited availability of places to plug in, competition from other vehicles, and consumer resistance to plugging in virtually every day) are likely to keep the number lower. The Maximum Practical rate depends on rapid technological progress, increased government support, and consumer acceptance. A more realistic penetration rate would result in 13 million PHEVs by 2030 out of about 300 million vehicles on the road, which still assumes that current levels of government support will continue for several decades.

5. PHEVs will have little impact on oil consumption before 2030 because there will not be enough of them in the fleet. More substantial reductions could be achieved by 2050. PHEV-10s will reduce oil consumption only slightly more than can be achieved by HEVs. A PHEV-10 is expected to use about 20 percent less gasoline than an equivalent HEV, saving about 70 gallons in 15,000 miles. Forty million PHEV-10s would save a total of about 0.2 million barrels of oil per day. The current light-duty vehicle fleet uses about 9 million barrels per day. PHEV-40s will consume about 55 percent less gasoline than equivalent HEVs, saving more than 200 gallons of gasoline per year per vehicle.

6. PHEV-10s will emit less carbon dioxide than non-hybrid vehicles, but save little relative to HEVs after accounting for emissions at the generating stations that supply the electric power. PHEV-40s are more effective than PHEV-10s, but the GHG benefits are small unless the grid is decarbonized with renewable energy, nuclear plants, or fossil fuel fired plants equipped with carbon capture and storage systems.

7. No major problems are likely to be encountered for several decades in supplying the power to charge PHEVs, as long as most vehicles are charged at night. Generation and transmission of electricity during off-peak hours should be adequate for many millions of PHEVs, although some distribution circuits may need upgrading if they are to serve clusters of PHEVs. Encouraging PHEV owners to charge their vehicles during off-peak hours will require both rate schedules that reward time-appropriate charging and equipment that can monitor—or even control—time of use.

8. A portfolio approach to research, development, demonstration, and, perhaps, market transition support is essential. It is not clear what technology or combination of technologies—batteries, hydrogen, or biofuels—will be most effective in reducing the nation's oil dependency to levels that may be necessary in the long run. It is clear, however, that a portfolio approach will enable the greatest reduction in oil use. Increasing the efficiency of conventional vehicles (including HEVs) beyond the current regulatory framework could reduce gasoline consumption by about 40 percent in 2050, compared to the Reference Case. Adding biofuels would reduce it another 20 percent. If PHEV-10s are also included at the Maximum Practical rate, gasoline consumption would be reduced an additional 7 percent, while PHEV-40s could reduce consumption by 23 percent. Employing HFCVs instead of PHEVs could eliminate gasoline use by the fleet.

References

Anderman, M. 2007. Gap Analysis for Li-ion Batteries for Automotive Applications. Advanced Automotive Batteries Conference 2007. Long Beach, California.

Bureau of the Census. 2008. American Housing Survey. Available at http://www.census.gov/hhes/www/housing/ahs/ahs07/tab2-25.pdf.

Carney, D. 2009. Electric Cars Poised to Give Auto Industry a Jolt. Available at www.msnbc.com. Accessed July 16, 2009.

DOE (Department of Energy). 2009a. Obama Administration Announces Availability of $3.9 Billion to Invest in Smart Grid Technologies and Electric Transmission Infrastructure. June 25. Available at http://www.energy.gov/news2009/7503.htm. Accessed December 1, 2009.

DOE. 2009b. Recovery Act Awards for Electric Drive Vehicle Battery and Component Manufacturing Initiative. August 5. Available at http://www1.eere.energy.gov/recovery/pdfs/battery_awardee_list.pdf. Accessed December 1, 2009.

DOE/EERE (Energy Efficiency and Renewable Energy). 2009. North American PHEV Demonstration. Available at http://www1.eere.energy.gov/vehiclesandfuels/avta/pdfs/phev/hymotion_prius_v2green_todate.pdf. Accessed January 26, 2010.

EIA (Energy Information Administration). 2008. Annual Energy Outlook 2008. Report No. DOE/EIA-0383(2008). Washington, D.C.: DOE, EIA.

EIA. 2009a. Annual Energy Outlook 2009 with Projections to 2030. Report No. DOE/EIA-0383(2009). Washington, D.C.: DOE, EIA. March.

EIA. 2009b. Annual Energy Review 2008. Report No. DOE/EIA-0384. Washington, D.C.: DOE, EIA. June.

Elgowainy, A., A. Burnham, M. Wang, J. Molburg, and A. Rousseau. 2009. Well-to-wheels Energy Use and Greenhouse Gas Emissions Analysis of Plug-in Hybrid Electric Vehicles. Argonne National Laboratory report ANL/ESD/09-2. Available at http://www.transportation.anl.gov/pdfs/TA/559.pdf. Accessed January 26, 2010.

EPRI (Electric Power Research Institute)/NRDC (Natural Resources Defense Council). 2007. Environmental Assessment of Plug-In Hybrid Electric Vehicles. Volume 1: Nationwide Greenhouse Gas Emissions. July. Available at http://my.epri.com/portal/server.pt?open=514&objID=223132&mode=2. Accessed December 1, 2009.

Francfort, J. 2009. Plug-in Hybrid Electric Vehicles (PHEV) Overview. Local Climate Energy Summit, Idaho National Laboratory, May.

Hensley, R., S. Knupfer, and D. Pinner. 2009. Electifying Cars: How Three Industries will Evolve, McKinsey and Co. Available at https://www.mckinseyquarterly.com/Electrifying_cars_How_three_industries_will_evolve_2370. June.

Howell, D. 2009. Annual Merit Review: Energy Storage R&D Overview. Washington, D.C.: DOE.

Howell, D., et al. 2009. Current Status of D.O.E.-Funded R&D Energy Storage for Automotive Applications. EVS24 International Battery, Hybrid, and Fuel Cell Electric Vehicle Symposium. Stavanger, Norway.

Kalhammer, F.R., B. Kopf, D. Swan, V. Roan, and M. Walsh. 2007. Status and Prospects for Zero Emissions Vehicle Technology. Sacramento, State of California Air Resources Board. Available at http://www.arb.ca.gov/msprog/zevprog/zevreview/zev_panel_report.pdf.

Kerr, R.A. 2008. Energy: World oil crunch looming? Science 322(5905): 1178-1179.

Kromer, M., and J. Heywood. 2007. Electric Powertrains: Opportunities and Challenges in the U.S. Light-Duty Vehicle Fleet. Publication LFEE 2007-02 RP. Cambridge, Mass.: Sloan Automotive Laboratory, Massachusetts Institute of Technology. May.

Morrow, K., D. Karner, and J. Francfort. 2008. Plug-in Hybrid Electric Vehicle Charging Infrastructure Review. INL/EXT-08-15058. Battelle Energy Alliance. November.

NAS-NAE-NRC (National Academy of Sciences-National Academy of Engineering-National Research Council). 2009. America's Energy Future: Technology and Transformation. Washington, D.C.: The National Academies Press.

Nelson, P.A., D.J. Santini, and J. Barnes. 2009. Factors Determining the Manufacturing Costs of Lithium-Ion Batteries for PHEVs. EVS24 International Battery, Hybrid and Fuel Cell Electric Vehicle Symposium, Stavanger, Norway, May 13-16, 2009.

NERC (North American Electric Reliability Corporation). 2007. Long-Term Reliability Assessment: 2007-2016. October.

NERC. 2008. Long-Term Reliability Assessment: 2008–2017. October.

NRC (National Research Council). 2008. Transitions to Alternative Energy Technologies—A Focus on Hydrogen. Washington, D.C.: The National Academies Press.

Pesaran, A., T. Market, H. Tataria, and D. Howell. 2007. Battery Requirements for Plug-in Hybrid Electric Vehicles: Analysis and Rational. 23rd International Electric Vehicle and Symposium and Exposition (EVS-23). Anaheim, California. Available at http://www.nrel.gov/vehiclesandfuels/energystorage/pdfs/42469.pdf.

PNNL (Pacific Northwest National Laboratory). 2007. Impacts Assessment of Plug-In Hybrid Vehicles on Electric Utilities and Regional U.S. Power Grids. Part 1: Technical Analysis. Available at http://www.ferc.gov/about/com-mem/wellinghoff/5-24-07-technical-analy-wellinghoff.pdf.

Plotkin, S., and Margaret Singh, Argonne National Laboratory. 2009. Multi Path Transportation Futures Study: Phase 2. Available at http://www.transportation.anl.gov/pdfs/TA/613.PDF.

Reuters. 2009. Saudi warns of "catastrophic" energy crunch, March 19.

Schlindler, J., et al. 2008. Crude Oil The Supply Outlook, Energy Watch Group, February.

Shiau, N., et al., 2009. Impact of Battery Weight and Charging Patterns on the Economic and Environmental Benefits of Plug-in Hybrid Electric Vehicles. Energy Policy 37. Available at http://www.cmu.edu/me/ddl/publications/2009-EP-Shiau-Samaras-Hauffe-Michalek-PHEV-Weight-Charging.pdf.

Simpson, A. 2006. Cost-Benefit Analysis of Plug-In Hybrid Electric Vehicle Technology. Conference paper NREL/CP-540-40485, presented at the 22nd International Battery, Hybrid and Fuel Cell Electric Vehicle Symposium and Exhibition (EVS-22). Yokohama, Japan. November. Available at http://www.nrel.gov/vehiclesandfuels/vsa/pdfs/40485.pdf.

Tikhonov, K., and V.R. Koch. 2009. Li-ion Battery Electrolytes Designed for a Wide Temperature Range. Woburn, Mass.: Covalent Associates Inc.

U.K. Industry Taskforce on Peak Oil and Energy Security. 2008. The Oil Crunch. Securing the UK's Energy Future. First report, October 29.

USABC (United States Advanced Battery Consortium LLC). 2009. Available at http://www.uscar.org/guest/article_view.php?articles_id=85.

Zhang, S.S., T.R. Jow, K. Amine, and G.L. Henricksen. 2002. $LiPF_6$-EC-EMC electrolyte for Li-ion battery. Journal of Power Sources 107(1): 18-23.

Appendixes

Appendix A

Committee Biographical Information

Michael P. Ramage, NAE, *chair*, is a retired vice president, ExxonMobil Research and Engineering Company. Previously he was executive vice president and chief technology officer, Mobil Oil Corporation. Dr. Ramage held a number of positions at Mobil, including research associate, manager of process research and development, general manager of exploration and producing research and technical service, vice president of engineering, and president of Mobil Technology Company. He has broad experience in many aspects of the petroleum and chemical industries. He has served on a number of university visiting committees and was a member of the Government University Industrial Research Roundtable. He was a director of the American Institute of Chemical Engineers and is a member of several professional organizations. Dr. Ramage chaired the recent National Research Council report *The Hydrogen Economy: Opportunities, Costs, Barriers, and Research Needs* (2004). He has served on the NAE Council. Dr. Ramage has B.S., M.S., Ph.D., and H.D.R. degrees in chemical engineering from Purdue University.

Rakesh Agrawal, NAE, is Winthrop E. Stone Distinguished Professor, School of Chemical Engineering, Purdue University. Previously, he was an Air Products Fellow at Air Products and Chemicals, Inc., where he worked from 1980 to 2004. A major thrust of his research is related to energy issues and includes novel processes for fabrication of low-cost solar cells, biomass and coal to liquid fuel conversion, hydrogen production from renewable sources, and energy systems analysis. His research interests further include basic and applied research in gas separations, process development, synthesis of distillation column configurations, adsorption and membrane separation processes, novel separation processes, gas liquefaction processes, cryogenics, and thermodynamics. He holds 116 U.S. and more than 500 foreign patents. These patents are used in over a hundred chemical plants with a capital expenditure in excess of a billion dollars. He has authored 66 technical papers and given many lectures and presentations. He chaired the Separations Division and the Chemical Technology Operating Council of the American Institute of Chemical Engineers (AIChE) and also a Gordon Conference on Separations. He was a member of the NRC Committee on Alternatives and Strategies for Future Hydrogen Production and Use. He is currently a member of the AIChE's Board of Directors and also its Energy Commission. He is also a member of the NRC Board on Energy and Environmental Systems (BEES). He has received several awards, including the J & E Hall Gold Medal from the Institute of Refrigeration (U.K.); Presidential Citation for Outstanding Achievement from the University of Delaware and from the AIChE; and the Gerhold Excellence in Industrial Gases Technology Institute Lecture and Chemical Engineering Practice awards. Dr. Agrawal received a B. Tech. from the Indian Institute of Technology, in Kanpur, India; an M.Ch.E. from the University of Delaware, and an Sc.D. in chemical engineering from the Massachusetts Institute of Technology.

David L. Bodde serves as a professor and senior fellow at Clemson University. There, he directs innovation and strategy at the Clemson's International Center for Automotive Research. Prior to joining Clemson University, Dr. Bodde held the Charles N. Kimball Chair in Technology and Innovation at the University of Missouri in Kansas City. Dr. Bodde serves on the board of directors of several energy and technology companies, including the Great Plains Energy and the Commerce Funds. His executive experience includes vice president, Midwest Research Institute; assistant director of the Congressional Budget Office; and deputy assistant secretary in the U.S. Department of Energy. Dr. Bodde frequently testifies before congressional committees. He was once a soldier and served in the Army in Vietnam. He has a doctorate in business administration from Harvard University, M.S. degrees in nuclear engineering (1972) and management (1973), and a B.S. from the U.S. Military Academy.

David Friedman, research director, Clean Vehicles Program, Union of Concerned Scientists (UCS), Washington, D.C. He is the author or coauthor of more than 30 technical papers and reports on advancements in conventional, fuel cell, and hybrid electric vehicles and alternative energy sources with an emphasis on clean and efficient technologies. Before joining UCS in 2001, he worked for the University of California-Davis (UC Davis) in the Fuel Cell Vehicle Modeling Program, developing simulation tools to evaluate fuel cell technology for automotive applications. He worked on the UC Davis FutureCar team to build a hybrid electric family car that doubled its fuel economy. He previously worked at Arthur D. Little researching fuel cell, battery electric, and hybrid electric vehicle technologies, as well as photovoltaics. He served as a member of the NRC Panel on the Benefits of Fuel Cell R&D of the Committee on Prospective Benefits of DOE's Energy Efficiency and Fossil Energy R&D Programs, Phase 1; on the Panel on Benefits of DOE's Light-Duty Hybrid Vehicle R&D Program; and as a member of the NRC Committee on National Tire Efficiency. He earned a bachelor's degree in mechanical engineering from Worcester Polytechnic Institute and is a doctoral candidate (2007) in transportation technology and policy at UC Davis.

Susan Fuhs is president, Conundrum Consulting. Previous positions include general manager, Astro Aerospace; general manager, GE Hybrid Power Generation Systems; director, New Ventures, Honeywell International; technology policy analyst, RAND; and project engineer, Advanced Applications, AlliedSignal Aerospace. Dr. Fuhs's technical and business experience has focused on overcoming barriers to the development and implementation of advanced technologies. Her experience with fuel cells includes developing fuel cell systems for stationary and transportation applications, including fuel cells for the Partnership for a New Generation of Vehicles; developing fuel cell marketing and business plans; and managing the solid oxide fuel cell subsidiary of General Electric Power Systems. She currently consults in strategic planning, new product development, business development, and technology roadmapping. She is a past board member, National Hydrogen Association, and past chairperson, Space Systems Technical Committee, American Institute of Aeronautics and Astronautics. She has a Ph.D. and M.S. in mechanical engineering and a B.S. in chemical engineering from the California Institute of Technology and an MBA from the Anderson School of the University of California, Los Angeles.

Judi Greenwald is the director of Innovative Solutions at the Pew Center on Global Climate Change. She oversees the Solutions program and develops mechanisms for learning about and promoting innovative solutions, including research, publications, Web-based information and databases, and workshops. Ms. Greenwald focuses on technological innovation, business solutions, and state and regional solutions. Ms. Greenwald has over 20 years of experience working on energy and environmental policy. Prior to coming to the Pew Center, she worked as a consultant, focusing on innovative approaches to solving environmental problems, including climate change. She also served as a senior advisor on the White House Climate Change Task Force. As a member of the professional staff of the U.S. Congress Energy and Commerce Committee, she worked on the 1990 Clean Air Act Amendments, the 1992 Energy Policy Act, and a number of other energy and environmental statutes. She was also a Congressional Fellow with then-Senate Majority Leader Robert C. Byrd, an environmental scientist with the U.S. Nuclear Regulatory Commission, and an environmental engineer and policy analyst at the Environmental Protection Agency (EPA). Ms. Greenwald has a B.S. in engineering, cum laude, from Princeton University and an M.A. in science, technology, and public policy from George Washington University. She has published papers on the future of water quality monitoring, worker and community adjustment to climate change policy, a multimedia approach to radon, environmental policies affecting the development of newer coal technologies, and the implications for air quality analysis of extended lifetimes for coal-fired boilers.

Robert L. Hirsch is senior energy advisor, Management Information Services, Inc. (MISI). Formerly he was senior energy program advisor at SAIC. His past positions include senior energy analyst with the RAND Corporation; executive advisor to the president of Advanced Power Technologies, Inc.; vice president, Washington Office, Electric Power Research Institute; vice president and manager, Research and Technical Services Department, ARCO Oil and Gas Company; chief executive officer of ARCO Power Technologies, a company that he founded; manager, Baytown Research and Development Division; general manager, Exploratory Research, Exxon Research and Engineering Company; assistant administrator for Solar, Geothermal, and Advanced Energy Systems (Presidential appointment); and director, Division of Magnetic Fusion Energy Research, U.S. Energy Research and Development Administration. He has served on numerous advisory committees, including as a member of the DOE Energy Research Advisory Board and a number of DOE national laboratory advisory boards. He has served on several NRC committees, including the one that wrote the report *Fuels To Drive Our Future* (1990), which examined the economics and technologies for producing transportation fuels from U.S. domestic resources; the Committee on Alternatives and Strategies for Future Hydrogen Production and Use; and was chairman of the Committee to Examine the Research Needs of the Advanced Extraction and Process Technology Program. He served as chairman of the Board on Energy and Environmental Systems and is a National Associate of the Academies. He brings expertise in a number of areas of science and technology and business related to energy production and consumption, research and development, and public policy. He received a Ph.D. in engineering and physics from the University of Illinois.

James R. Katzer, NAE, is an independent consultant. He has recently been a visiting scholar at MIT working on an MIT study *The Future of Coal in a Carbon Constrained World.* Prior to that he was manager of strategic planning and program analysis for ExxonMobil Research and Engineering Company, where he was responsible for technology-planning and analysis activities. Before that he was vice president, Technology, Mobil Oil Corporation, with primary responsibilities for ensuring Mobil Oil's overall technical health, developing forward-looking technology scenarios, identifying and analyzing technology and environmental developments and trends, identifying future threats and opportunities and strategies to deal with them. Dr. Katzer joined the Central Research Laboratory of the Mobil Oil Corporation in 1981, later becoming manager of process research and technical service and vice president of planning and finance for the Mobil Research and Development Corporation. Before joining Mobil he was a professor on the chemical engineering faculty at the University of Delaware and the first director of the Center for Catalytic Science and Technology there. He recently served on the NRC Committee on Alternatives to Indian Point that evaluated various energy supply and end-use technologies as potential replacements for the Indian Point nuclear power plants. Dr. Katzer has more than 80 publications in technical journals, holds several patents, and co-authored and edited several books. He received a B.S. from Iowa State and a Ph.D. in chemical engineering from MIT.

Gene Nemanich is the retired vice president of Hydrogen Systems for Chevron Technology Ventures, where he was responsible for hydrogen supply and developing and commercializing new hydrogen technologies. He has 32 years of experience with integrated oil companies, including Exxon, Cities Service, Texaco, and Chevron. He has also worked in the areas of refining, clean coal technology, oil supply and trading, and research leading to the development of new hydrogen systems. He represented Texaco in the California Fuel Cell Partnership in 2000-2001 and was a director of Texaco Ovonic Hydrogen Systems LLC, a joint venture with Energy Conversion Devices to commercialize metal hydride hydrogen storage systems. He was one of seven industry leaders that helped prepare the DOE-sponsored Hydrogen Roadmap and has served as chairman of the National Hydrogen Association. He has a B.S. in chemical engineering from the University of Illinois and an MBA from the University of Houston.

Joan Ogden is professor of environmental science and policy and an energy policy analyst at the Institute of Transportation Studies, University of California, Davis. Previous to this, she held a number of positions at various research institutions, including research scientist, Center for Energy and Environmental Studies, Princeton University. Most of her work has involved technical and economic assessments of new energy technologies, including renewable fuels, the use of hydrogen as an energy carrier, and applications of fuel cell technology in transportation. Particular areas of interest are production of renewable fuels, the use of hydrogen as an energy carrier, and applications of fuel cells in transportation and stationary power production. Over the past decade, Dr. Ogden has carried out a series of assessments of fuel cell vehicles and hydrogen refueling infrastructure. For 2 years, she served as chairman of the Solar Fuels and Transportation Division of the American Solar Energy Society. She has worked with the H2A, a group of hydrogen analysts convened by the Department of Energy to develop a consistent framework for analyzing hydrogen systems and, in 2005 and 2006, received R&D Excellence awards from the DOE for her work with H2A. In 2004, Dr. Ogden served on the governor of California's advisory panel developing a blueprint plan for the proposed California Hydrogen Highway Network. Dr. Ogden has published over 100 technical articles on energy topics, including the book *Solar Hydrogen.* She received her Ph.D. in physics from the University of Maryland and a B.S. in mathematics, University of Illinois, Champaign-Urbana.

Lawrence T. Papay, NAE, is currently a consultant with a variety of clients in electric power and other energy areas. His previous positions include senior vice president for the Integrated Solutions Sector, SAIC; and senior vice president and general manager of Bechtel Technology and Consulting. He also held several positions at Southern California Edison, including senior vice president, vice president, general superintendent, and director of R&D, with responsibilities for areas including bulk power generation, system planning, nuclear power, environmental operations, and development of the organization and plans for the company's R&D efforts. His professional affiliations have included the EPRI Research Advisory Committee, the Atomic Industrial Forum, the DOE Energy Research Advisory Board, and the Renewable Energy Institute. He is a member of the National Academy of Engineering and the National Science Foundation's Industrial Panel on Science and Technology. His expertise and knowledge ranges across a wide variety of electric system technologies, from production, to transmission and distribution, utility management and systems, and end-use technologies. He received a B.S. in physics from Fordham University and an S.M. and Sc.D. in nuclear engineering from MIT.

Ian Parry is a senior fellow at Resources for the Future. Previous positions include adjunct professor, Department of Economics, Georgetown University; research fellow, U.S. Department of Agriculture; professor, Center for Economic Research and Graduate Education (Prague); and lecturer,

Department of Economics, Australian National University. Dr. Parry's research focuses primarily on environmental, transportation, tax, and public health policies. His recent work has analyzed gasoline taxes, fuel economy standards, transit subsidies, alcohol taxes, policies to reduce traffic congestion and accidents, environmental tax shifts, the role of technology policy in environmental protection, the incidence of pollution control policies, and the interactions between regulatory policies and the broader tax system. He received a Ph.D. in economics from the University of Chicago, an M.A. in economics from Warwick University, and a B.A. in economics from the University of Sheffield.

William F. Powers, NAE, is retired vice president, research, Ford Motor Company. His approximately 20 years at Ford included positions as director, Vehicle, Powertrain and Systems Research; director, Product and Manufacturing Systems; program manager, Specialty Car Programs; and executive director, Ford Research Laboratory and Information Technology. Prior positions also include professor, Department of Aerospace Engineering, University of Michigan, during which time he consulted with NASA, Northrop, Caterpillar, and Ford; research engineer, University of Texas; and mathematician and aerospace engineer, NASA Marshall Space Flight Center. Dr. Powers is a fellow at the Institute of Electrical and Electronics Engineers, the Society of Automotive Engineers, the American Society of Mechanical Engineers, and the International Federation of Automatic Control. He is a foreign member of the Royal Swedish Academy of Engineering Sciences. He has extensive expertise in advanced research and development of automotive technology. He is a member of the National Academies' Board on Energy and Environmental Systems, and recently served on the Committee on Alternatives and Strategies for Future Hydrogen Production and Use. He has a B.S. in aerospace engineering, University of Florida, and a Ph.D. in engineering mechanics, University of Texas-Austin.

Edward S. Rubin is the Alumni Professor of Environmental Engineering and Science at Carnegie Mellon University (CMU). He holds joint appointments in the Departments of Engineering and Public Policy and Mechanical Engineering and is the founding director of CMU's Environmental Institute and Center for Energy and Environmental Studies. His teaching and research interests at CMU are in environmental control, energy utilization, and technology-policy interactions, with a particular focus on coal-based systems. His expertise includes modeling and assessment of energy and environmental systems with applications to electric power generation technologies, energy use, and emission control systems; global climate change policy issues; carbon sequestration and management; and environmental technology innovation and its relation to government policies. He has served as a member of numerous technical and advisory committees, including to the U.S. Environmental Protection Agency, the U.S. Department of Energy, the Intergovernmental Panel on Climate Change, and the National Academy of Sciences/National Research Council. He is a past chairman of the Environmental Control Division of the American Society of Mechanical Engineers. He earned a B.E. in mechanical engineering from the City College of New York and an M.S. and a Ph.D. in mechanical engineering from Stanford University.

Robert W. Shaw, Jr., is president of Aretê Corporation, the manager of the Micro-Generation Technology Fund, LLC, and the five Utech venture funds. He has over 20 years of experience in the venture capital industry and is a leader in developing modular/dispersed generation, renewable energy generation, hydrogen energy systems, and specialty materials. He previously held the position of senior vice president and was a member of the board of directors of Booz, Allen & Hamilton, where he was a founder of the firm's Energy Division, which provided management and technical consulting services to utilities and energy companies. He also held research positions at Bell Laboratories and Cavendish Laboratory directed at the electronic and structural properties of materials. Dr. Shaw served for 11 years as director and chairman of Distributed Energy Systems Corporation (DESC) and for 5 years as director and chairman of CTP Hydrogen Corporation. He has been a director of H2Gen Innovations, Inc., since 2001. He has served as a member of the NRC Board on Energy and Environmental Systems and the Panel on Benefits of DOE's Fuel Cell R&D Program. He is also a member of DOE's Hydrogen Technology Advisory Committee. He has a Ph.D. in applied physics from Stanford University, an M.S.E.E. from Cornell University, and an M.P.A. in organization design from American University.

Tony Wu is principal research engineer and project manager at the DOE's National Carbon Capture Center, managed by Southern Company. He is responsible for CO_2 capture technology assessment and coordinating testing and demonstration of promising technologies at the center. He also served on the FutureGen Technical Committee from 2008 to 2009 for the development of the DOE-sponsored zero emission coal-fueled IGCC power plant in Mattoon, Illinois. Prior to the current position, he was responsible for multiple research areas at Southern Company, including distributed energy resources, hydrogen, electric transportation, and energy storage programs. He has more than 20 years of combined experience in fundamental research, technology assessment and development, product testing and validation, and project management. His technical expertise is in chemical, material and electrochemical behavior of various power generation and energy storage systems such as ultracapacitor, battery and fuel cell technologies. Previous positions include staff technology engineer, Energizer Power Systems, and technology engineer, Gates Energy Products. He has a B.S. in chemical engineering from Tamkang University (Taiwan) and an M.S. in chemical engineering from Auburn University.

Appendix B

Presentations and Committee Meetings

FIRST COMMITTEE MEETING

May 18, 2009, Washington, D.C.

David Vieau, A123 Systems
Mark Verbrugge, General Motors
Bill Reinert and Shinichi Abe, Toyota
Dick Cromie and Bob Graham, Southern California Edison
Sandy Thomas, H2Gen

SECOND COMMITTEE MEETING

June 18, 2009, Washington, D.C.

Phil Patterson, U.S. Department of Energy
Jake Ward, U.S. Department of Energy
Dave Howell, U.S. Department of Energy
Tien Nguyen, U.S. Department of Energy
Michael Wang, Argonne National Laboratory

THIRD COMMITTEE MEETING

September 1-2, 2009, Washington, D.C.

No open sessions were held during this meeting.

Appendix C

Scenarios

Chapter 4 of this report compares scenarios for light-duty vehicles between 2010 and 2050. This appendix provides details of how that analysis was performed. It also analyzes the transition costs to achieve cost-effectiveness. Finally, it provides more detail on the decarbonized grid discussed in Chapter 4.

SCENARIO ANALYSIS

The first three cases, which do not include plug-in hybrid electric vehicles (PHEVs), are taken directly from the 2008 Hydrogen Report. They provide a point of comparison for the PHEV cases. They also allow us to analyze portfolio cases, where a strategy of introducing PHEVs is combined with improving efficiency in the gasoline ICEVs and HEVs and with the introduction of biofuels.

Hydrogen Report Cases

- *Reference Case (same as the 2008 Hydrogen Report Reference Case).* Gasoline internal combustion engine vehicles (ICEVs) continue to dominate the light-duty sector (Figure C.1). Gasoline HEVs gain about 10 percent fleet share by 2050. The fuel consumption of ICEV and HEV vehicles follows projections from the EIA Annual Energy Outlook 2008, meeting CAFE standards by 2020, with only modest improvements in fuel economy beyond this time (Figure C.2).
- *ICEV Efficiency Case (2008 Hydrogen Report Case 2).* Improvements in internal combustion engine technology are implemented, and HEVs comprise 60 percent of the fleet by 2050 (Figure C.3). Fuel economy increases for both ICEVs and HEVs (Figure C.4).
- *Biofuels Intensive Case (2008 Hydrogen Report Case 3).* Biofuels are introduced at a rapid rate. Over time, lower carbon biofuel supply is implemented (Figure C.5).

PHEV Cases

- *PHEV Case 1.* PHEVs introduced according to Figure 4.1 (Chapter 4 in this report); total vehicles remain at Reference Case levels.
- *PHEV + ICEV Efficiency (PHEV Case 2).* Same as PHEV Case 1, but gasoline ICEVs and HEVs improve according to ICEV Efficiency Case (Hydrogen Report Case 2). Vehicle mix is shown in Figure C.6.
- *PHEV + ICEV Efficiency + Biofuels (PHEV Case 3).* Same as PHEV Case 2, but biofuels are rapidly introduced, replacing some of the fuel used by ICEVs and HEVs. Vehicle mix is shown in Figure C.6.

ESTIMATING PHEV PERFORMANCE

As illustrated in Figure C.7, while the battery is above a minimum state of charge (SOC), the PHEV operates in a charge-depleting (CD) mode, in which it draws down the onboard battery to meet vehicle power demands. Once it reaches this minimum SOC, the vehicle switches to charge-sustaining (CS) mode, which is functionally equivalent to conventional HEV operation. During this mode, the vehicle maintains the SOC within a limited operating envelope, using stored battery energy and capturing regenerative braking energy to optimize ICE operation.

For vehicles with a single source of stored energy, such as gasoline, hydrogen, or electric battery, modeling the energy consumption is fairly straightforward once the influencing factors (vehicle weight, frontal area, aerodynamic drag, rolling resistance, engine and drive-train component performance and efficiency, and drive cycle) are specified.

For plug-in hybrid vehicles, however, there are two sources of stored energy onboard, gasoline and electricity, adding complexity to the energy-modeling task. The model must include estimates of the fraction of vehicle miles traveled (VMT) on electricity and the VMT on gasoline and how

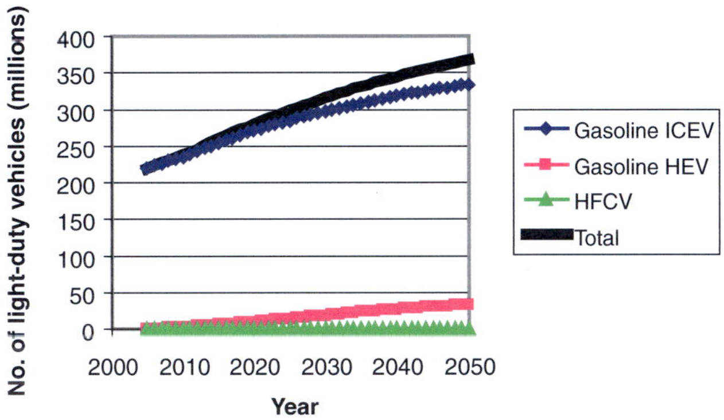

FIGURE C.1 Number of vehicles in the Hydrogen Report Reference Case. SOURCE: NRC, 2008.

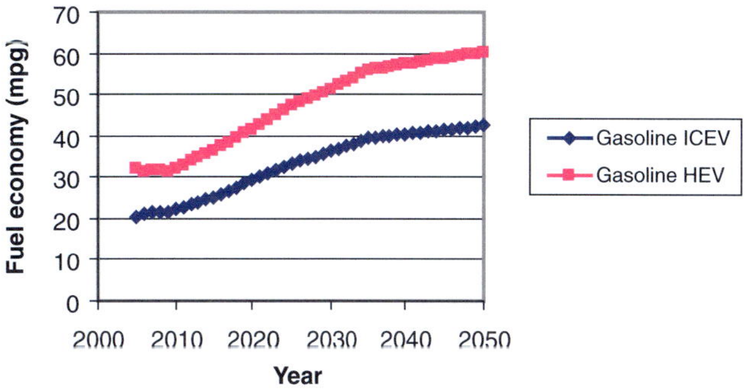

FIGURE C.4 Fuel economy for the ICEV Efficiency Case (Hydrogen Report Case 2). SOURCE: NRC, 2008.

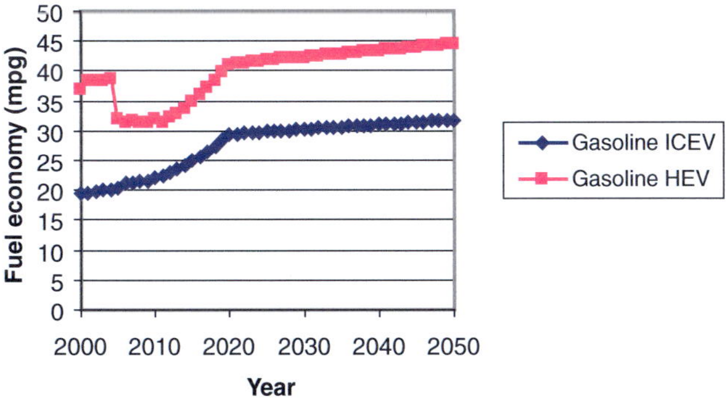

FIGURE C.2 Fuel economy for vehicles in the Hydrogen Report Reference Case. SOURCE: NRC, 2008.

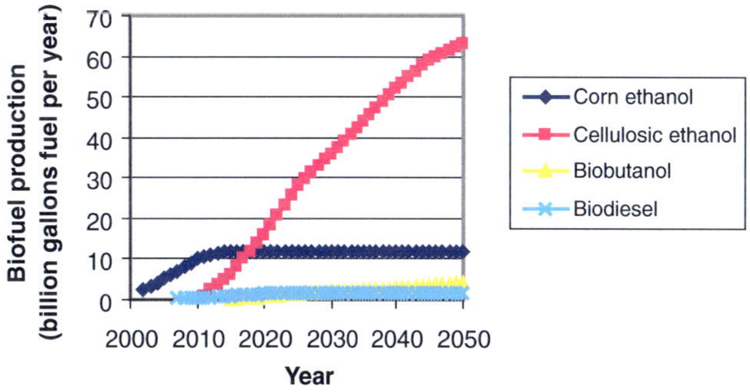

FIGURE C.5 Biofuel supply for the Biofuels-Intensive Case (Hydrogen Report Case 3). SOURCE: NRC, 2008.

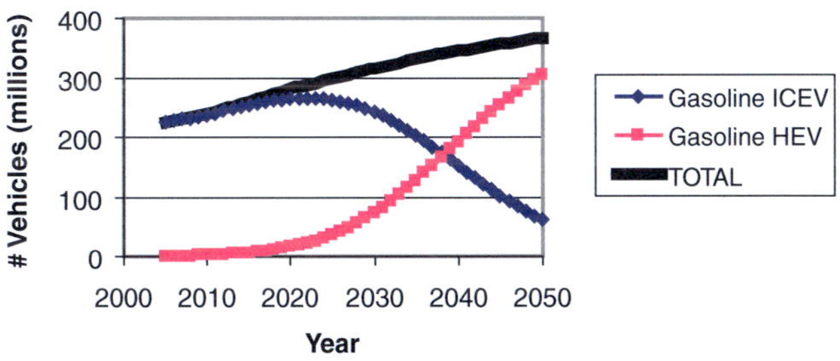

FIGURE C.3 Number of vehicles in the ICEV Efficiency Case (Hydrogen Report Case 2). SOURCE: NRC, 2008.

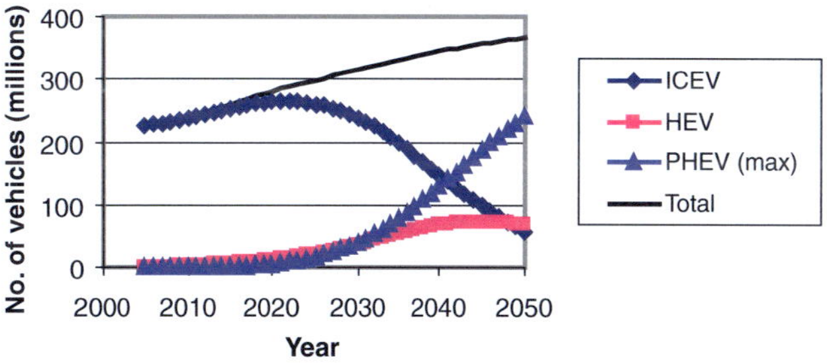

FIGURE C.6 Numbers of light-duty vehicles for portfolio approach, where PHEVs are combined with efficient ICEVs and HEVs.

much electricity and fuel are consumed over a drive cycle, both of which are influenced by three factors:

- *The size of the battery.* The larger the PHEV battery, the greater the fraction of the car's energy use that can be provided by electricity. Battery size is sometimes expressed as all-electric range (AER), the distance that could be traveled on just the battery if the car is operated in CD mode without using the engine.
- *Pattern of driving.* The fraction of miles traveled on electricity can also vary, depending on the driver's pattern of trips. If the driver takes only short trips (less than the all-electric range of the battery), all the miles could all be traveled on electricity. For longer trips, the driver will deplete the battery and will have to use the engine.
- *Control strategy of the PHEV when driven in CD mode.* Some PHEVs (the PHEV-40 in this report) use an all-electric strategy, where the battery is depleted to a minimum SOC.

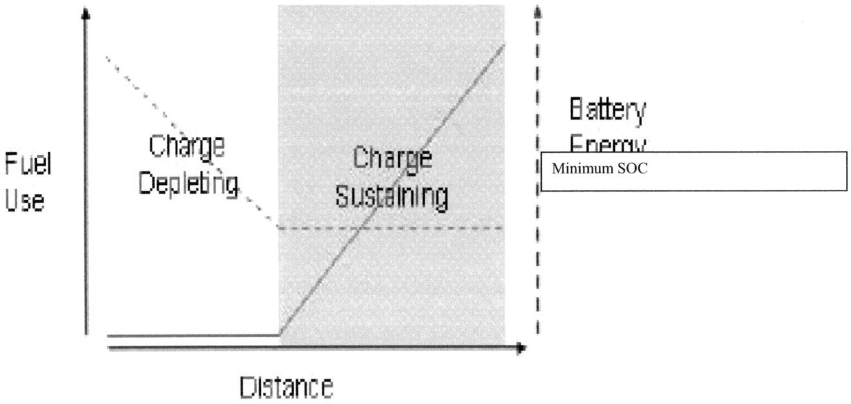

FIGURE C.7 PHEV operating modes. SOURCE: Kromer and Heywood, 2007.

At this point, the engine is turned on and the vehicle operates in CS mode, similar to a gasoline hybrid. Other PHEVs (the PHEV-10) use a "blended" strategy, where the engine is engaged when additional power is needed for acceleration or hill climbing as well as when the battery is discharged.

Vehicle simulation models were not used in this study. However, several recent studies have simulated a range of vehicles on a self-consistent basis, including gasoline ICEVs, HEVs, PHEVs, EVS, and HFCVs (Kromer and Heywood, 2007; Elgowainy et al., 2009; Simpson, 2006; Plotkin and Singh, 2009). These studies employ varying assumptions about PHEV design and control strategies.

To span the range of control strategies, the committee modeled a PHEV-40 with an all-electric drive strategy and a PHEV-10 with a blended strategy. Both PHEVs are midsize sedans with 100 kW power output. The committee drew on the results of the referenced studies to approximate the performance of the PHEVs modeled. This was accomplished in four steps.

Step 1. Estimate Fraction of Miles Driven in CD and CS Mode

The committee used a chart similar to Figure C.8 which estimates the utility factor—the fraction of miles that could be traveled on electricity in the United States—as a function of the PHEV's all-electric range, or battery size. For a PHEV-10, 23 percent of the nation's miles traveled could be on electricity. For a PHEV-40, the utility factor is 63 percent.

Step 2. Estimate PHEV Gasoline and Electricity Use over Drive Cycle

The committee took the energy-use values for PHEVs in CD and CS modes from the referenced reports. The energy-use values were then combined with the estimated fraction of miles spent in CD and CS modes from Step 1 to estimate electricity and fuel use over the whole drive cycle.

Figure C.9 illustrates the energy consumption of gasoline and electricity over the combined FTP/HWFET drive cycle for various types of advanced hybrid and plug-in hybrid vehicles. As battery size increases, gasoline consumption falls and electricity increases. The overall energy efficiency of the vehicle is higher with larger batteries.

Step 3. Estimate Energy Consumption for All-Electric and Blended Vehicles

PHEV energy use over a drive cycle depends on the degree of blending assumed during CD mode. For an all-electric strategy, petroleum consumption over a drive cycle is lower than for a blended strategy. This is illustrated in Figure C.10.

Electricity use is about the same for various degrees of blending, but gasoline use increases at higher blending ratios. Blended-26 percent represents the maximum possible blending. Blended-55 percent represents all-electric operation.

Only one study (Kromer and Heywood, 2007) evaluated both blended and all-electric-range operation, and the committee used that study for estimating PHEV energy use. Although a PHEV-40 was not specifically evaluated in the study, linear interpolation between PHEV-30 and PHEV-60 results provided estimated energy use for PHEV-40s.

PHEV gasoline and electricity energy use are expressed as fractions of the gasoline energy used in an HEV, as shown in Table C.1. These ratios put PHEV energy use on the same basis as the 2008 Hydrogen Report.

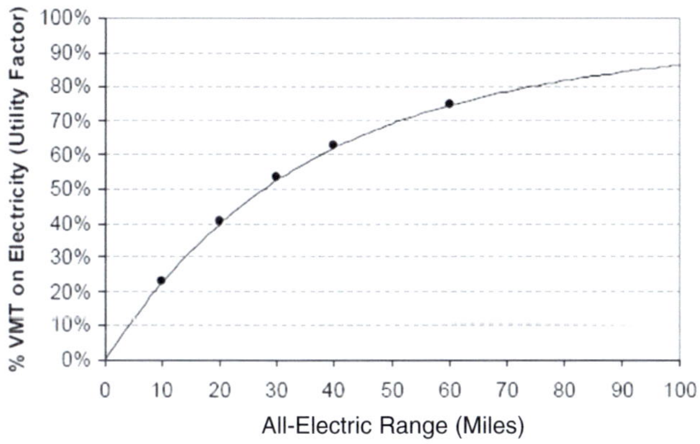

FIGURE C.8 National VMT fraction available for substitution by a PHEV using 100 percent electric charge-depleting mode.
SOURCE: Elgowainy et al., 2009.

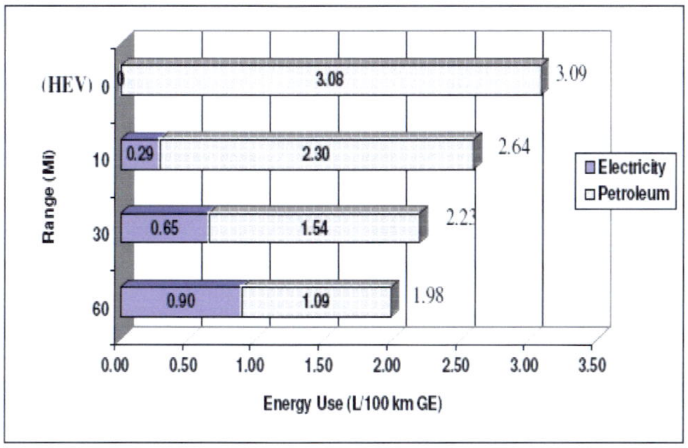

FIGURE C.9 Tank-to-wheels energy use in advanced vehicles, assuming 44 percent blending during charge-depleting operation.
SOURCE: Kromer and Heywood, 2007.

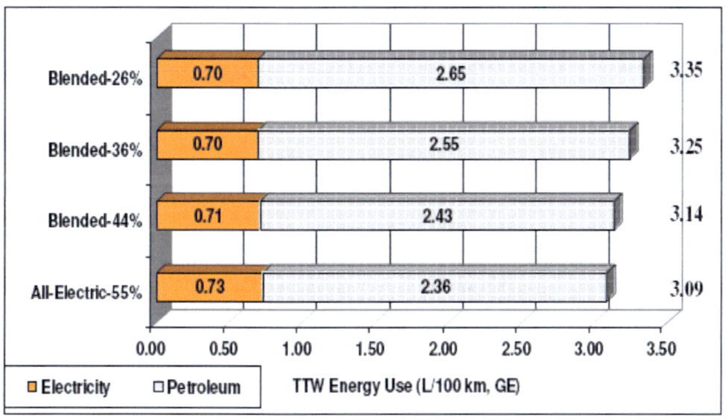

FIGURE C.10 Energy consumption in a PHEV-30 as electricity and gasoline for different blending strategies in CD mode.
SOURCE: Kromer and Heywood, 2007.

Step 4. Estimate PHEV Gasoline and Electricity Use over Time

The committee reasoned that PHEV engine and vehicle technologies (e.g., aerodynamics and rolling resistance) would improve at the same rate as fuel economy technologies in the ICEV Efficiency Case: 2.7 percent per year from 2010 to 2025; 1.5 percent per year from 2026 to 2035; and 0.5 percent per year from 2036 to 2050. Combining these improvement rates and the derived energy-use ratios in Table C.1, the committee then developed assumed values for gasoline and electricity use vs. time for the PHEV-10 and PHEV-40 from 2010 to 2050.

Figure C.11 shows the resulting gasoline use for PHEV-10 and PHEV-40 vehicles for the Optimistic technology case. Gasoline ICEV and HEV gasoline use in the Reference Case and high-efficiency cases are shown for comparison. Figure C.12 shows the estimated electricity use for both the PHEV-10 and the PHEV-40.

TABLE C.1 Ratio of Energy Use in PHEVs Compared to Energy Use in Gasoline HEVs

	Energy Use in PHEVs vs. Gasoline Use in HEVs	
Ratio	PHEV-10 (Blended CD)	PHEV-40 (All-Electric CD)
Gasoline use in PHEVs: gasoline use in HEVs	0.81	0.45
Electric energy use in PHEVs: gasoline energy use in HEVs	0.09	0.24

NOTE: The PHEV-10 is assumed to operate in blended mode and the PHEV-40 in all-electric mode during CD operation. As a check, the committee also calculated ratios for PHEV electrical energy use and gasoline use as compared to a hybrid vehicle for two other PHEV modeling studies (Elgowainy et al., 2009; Simpson, 2006). The results were broadly similar for the PHEV-10 (gasoline use was 85-88 percent of HEV gasoline use and electricity energy use was 4-5 percent of HEV gasoline energy use). For the PHEV-40, these two studies estimated gasoline consumption ratios of 55-60 percent, and electricity use 12-15 percent (higher gasoline use and lower electricity use than Kromer and Heywood). Part of the difference may be because Elgowainy et al. (2009) and Simpson (2006) simulated only blended-mode CD operation, while Kromer and Heywood (2007) considered all-electric mode.

SOURCE: Adapted from Kromer and Heywood (2007).

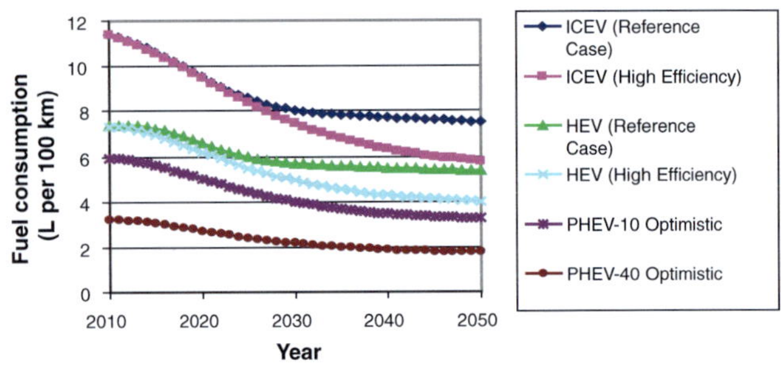

FIGURE C.11 Estimated on-road, fleet-average gasoline consumption for ICEVs, HEVs, and PHEVs in this study. Electricity use in PHEVs not included.

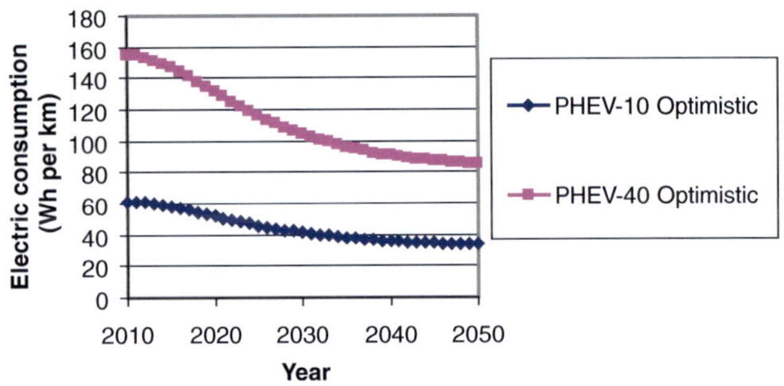

FIGURE C.12 Estimated fleet-average electricity use over drive cycle for PHEVs in this study.

TRANSITION COST ANALYSIS

A transition cash flow analysis was conducted to determine the investment costs required for PHEVs to reach cost competitiveness with Reference Case gasoline vehicles. For each year, the committee estimated the incremental cost of buying PHEVs instead of gasoline reference vehicles. The incremental investment for vehicles is (Reference vehicle price – PHEV price) times the number of PHEVs sold each year. Then the committee estimated the annual cost of fuel for all the PHEVs in the fleet and the cost of fuel for an equal number of gasoline reference vehicles. The breakeven is the year when annual fuel cost savings balance annual purchase cost differences. All cases assume that charging electricity costs 8 cents per kWh and that gasoline prices, as in the hydrogen study, increase from \$2.70 per gallon in 2010 to \$4.00 per gallon in 2050 (see Figure 4.9).

Results are shown in Figures C.13 through C.16 for PHEV-10s and PHEV-40s. "Maximum Practical" is the market penetration rate with Optimistic technical progress and "Probable" is the market penetration rate with Probable technical progress. In addition, a mixed case, where 30 percent of the market is captured by PHEV-40s and 70 percent by PHEV-10s, is also included (Figures C.17 and C.18). These figures supplement the results presented in Chapter 4 (Table 4.3).

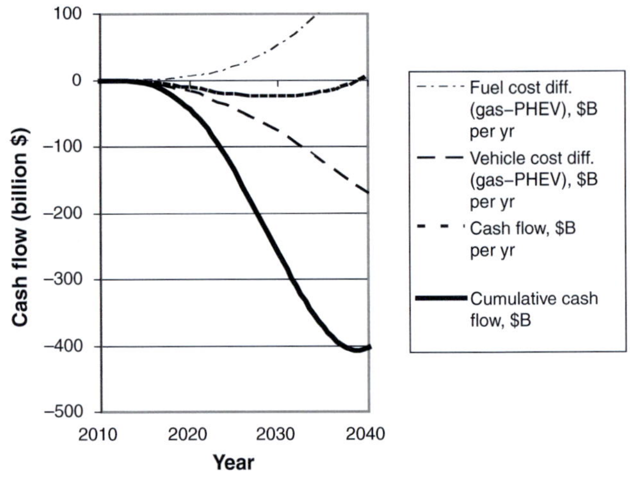

FIGURE C.13 Cash flow analysis for PHEV-40, Maximum Practical case, Optimistic technical assumptions. The break-even year is 2040, and the buydown cost is \$408 billion.

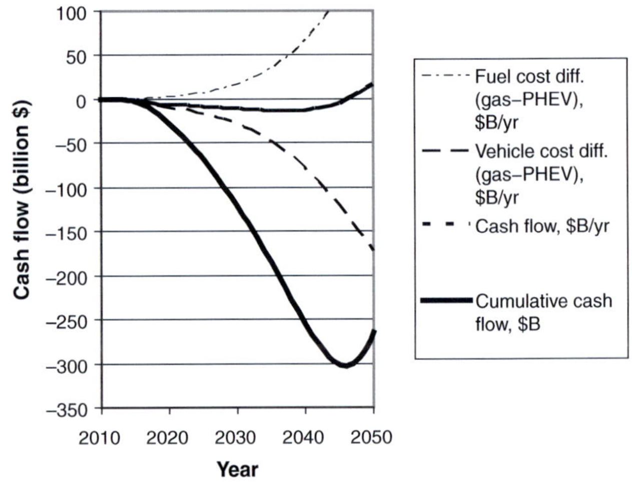

FIGURE C.14 Cash flow analysis for PHEV-40, Probable case, Probable technical assumptions. The break-even year is 2047, and the buydown cost is \$303 billion.

SENSITIVITY STUDIES

The sensitivity of the transition analysis was explored for four key parameters: the price of electricity, the price of gasoline, and the incremental costs of the PHEV-10 and the PHEV-40 relative to those of a reference vehicle. Base case values are shown in Table C.2. Each variable is normalized to the base case value in Table C.3, which allows the sensitivity results to be plotted on the same graph. Results for the break-even year and buydown cost for the two PHEVs are shown in Figures C.19 through C.22.

The buydown cost and break-even year for the PHEV-10 are not very sensitive to electricity prices, because most of

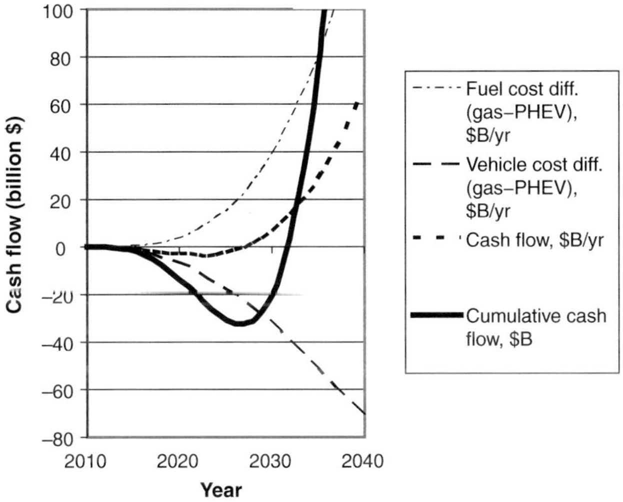

FIGURE C.15 Cash flow analysis for PHEV-10, Maximum Practical case, Optimistic technical assumptions. The break-even year is 2028, and the buydown cost is $33 billion.

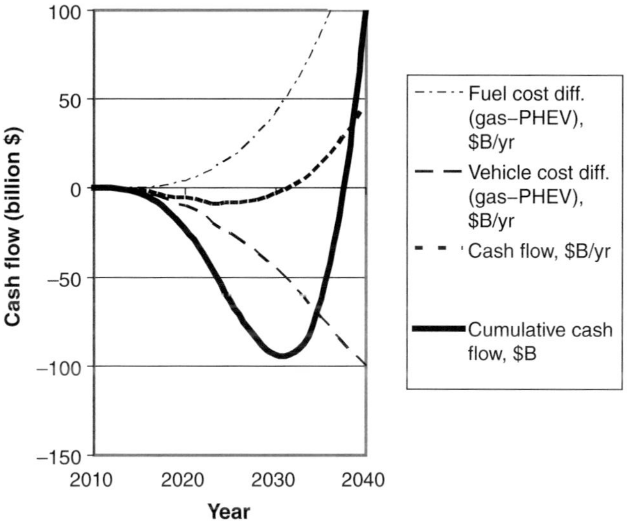

FIGURE C.17 Cash flow analysis for mixed case (70 percent PHEV-10s and 30 percent PHEV-40s), Maximum Practical case, Optimistic technical assumptions. The break-even year is 2032, and the buydown cost is $94 billion.

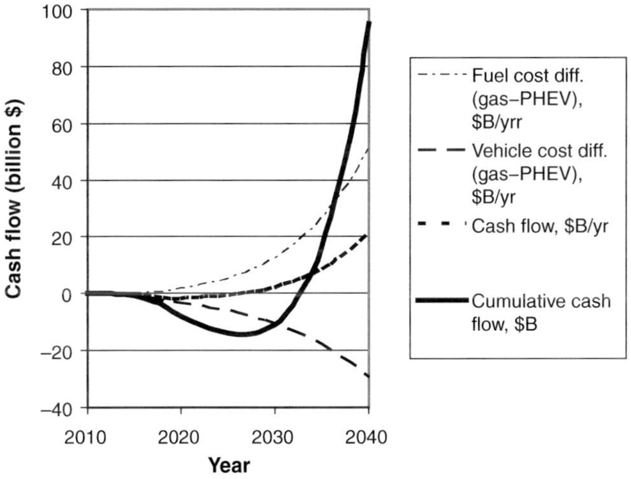

FIGURE C.16 Cash flow analysis for PHEV-10, Probable case, Probable technical assumptions. The break-even year is 2028, and the buydown cost is $15 billion.

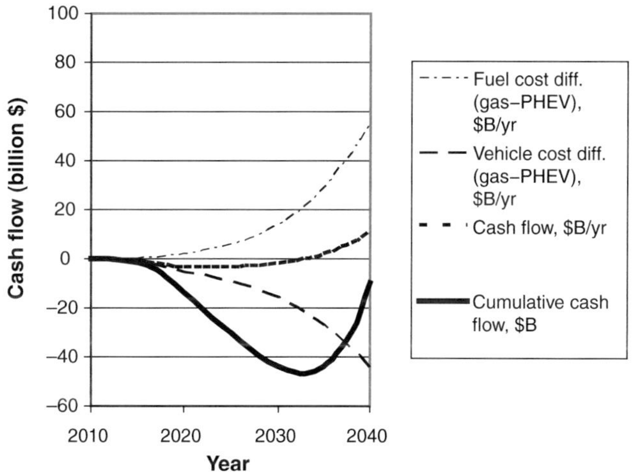

FIGURE C.18 Cash flow analysis for mixed case (70 percent PHEV-10s and 30 percent PHEV-40s), Probable Case, Probable technical assumptions. The break-even year is 2034, and the buydown cost is $47 billion.

the fuel used by the PHEV-10 is gasoline. The PHEV-40 results show a higher sensitivity to electricity price, as these vehicles travel over half their miles on electricity. Even if the electricity price was 12 cents per kWh instead of the base case (8 cents per kWh), breakeven for the PHEV-40 would be delayed only about 2 years.

The results for both PHEV-10 and PHEV-40 are sensitive to the assumed oil price. If oil prices rose 50 percent compared to our base case (price of $120-$180/bbl or $4-$6/gallon gasoline in the timeframe 2010-2030), the PHEV-40 would break even in 2029 (instead of 2040), and buydown costs would be reduced to about $100 billion (from $400 billion).

Finally, the break-even year and the buydown cost are sensitive to the assumed vehicle price and the rate of learning. In the low-cost case, the committee assumes that DOE goals are met by 2020. This implies an earlier break-even year and a much lower buydown cost for both the PHEV-10

TABLE C.2 Input Variables for Sensitivity Study

Parameter	Low	Base	High
Electricity price, $ per kWh	0.06	0.08	0.15
Gasoline Price $ per gal[a]	0.5 × Base	1.0 × Base	2.0 × Base
Vehicle Incremental retail price, $[b]	DOE Goal (2020)	Optimistic	Probable
PHEV-10[c]	4,500	7,700 (2010) 5,100 (2030)	8,800 (2010) 5,700 (2030)
PHEV-40[d]	7,600	19,800 (2010) 12,300 (2030)	25,500 (2010) 15,500 (2030)

[a]DOE's High Price Case (EIA, 2008, *Annual Energy Outlook*). See Figure 4.9, which shows gasoline prices ranging from $2.75 to $4.00 per gallon from 2010 to 2050. Corresponds to oil at $80 to $120 per barrel (2010-2030).

[b]See Table 4.2.

[c]OEM cost of battery, $ per usable kWh: 2020, $500 (DOE goal); 2030, base, $720; 2030, high, $950.

[d]OEM cost of battery, $ per usable kWh: 2020, $300 (DOE goal); 2030, base, $720; 2030, high, $1000.

TABLE C.3 Range of Inputs Normalized to Base Value (divide values in Table C.2 by base value)

Variable	Low	Base	High
Electricity Price $ per kWh	0.75	1	1.875
Gasoline Price $ per gal[a]	0.5 × Base	1.0 × Base	2.0 × Base
Vehicle Incremental retail price, $[b]	DOE Goal	Optimistic	Probable
PHEV-10[c]	Base 0.87	1	Base 1.13
PHEV-40[d]	Base 0.62		Base 1.25

[a]Base is DOE's High Price Case (EIA, 2008, *Annual Energy Outlook*). See Figure 4.9, which shows gasoline prices ranging from $2.75 to $4.00 per gallon from 2010 to 2050, corresponding to oil at $80 to $120 per barrel (2010-2030).

[b]See Table 4.2.

[c]OEM cost of battery, $ per usable kWh: 2020, $500 (DOE goal); 2030, base, $720; 2030, high, $950.

[d]OEM cost of battery, $ per usable kWh: 2020, $300 (DOE goal); 2030, base, $720; 2030, high, $1000.

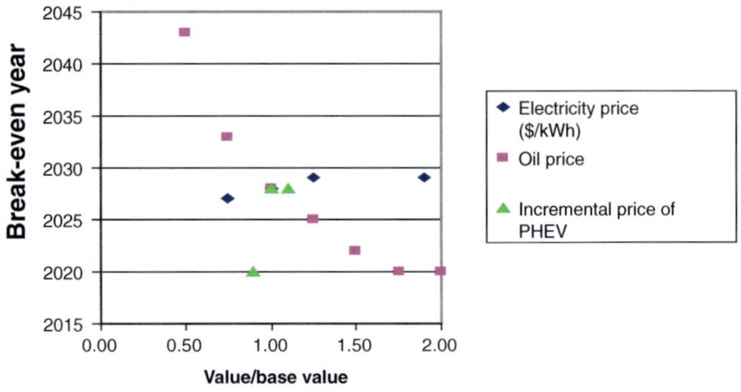

FIGURE C.19 PHEV-10: Sensitivity of break-even year to changes in input variables.

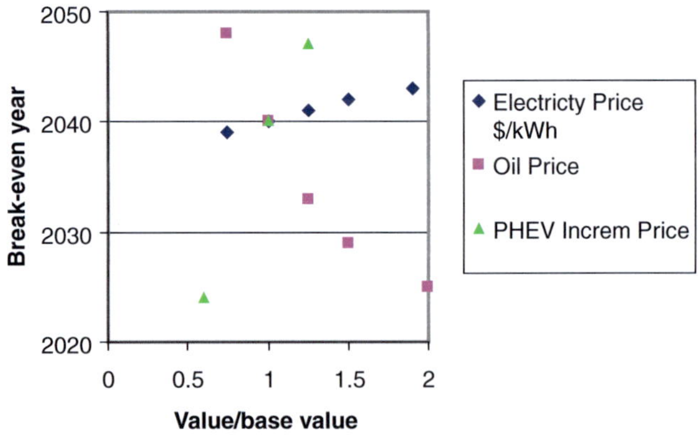

FIGURE C.20 PHEV-40: Sensitivity of break-even year to changes in input variables.

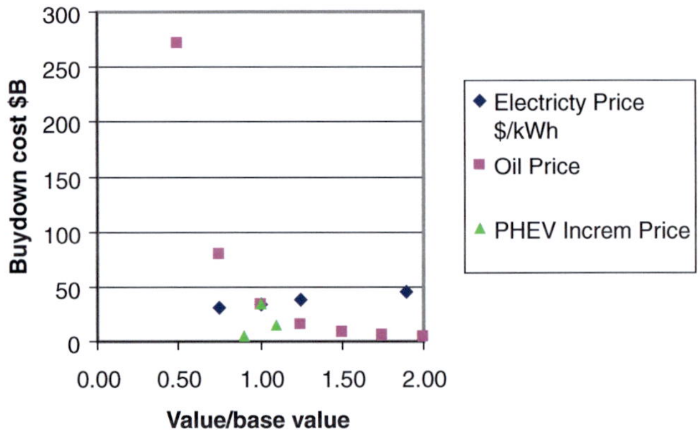

FIGURE C.21 PHEV-10: Sensitivity of buydown cost to changes in input variables.

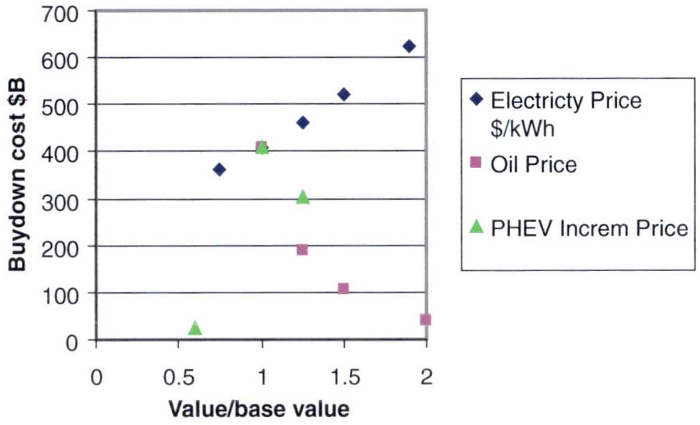

FIGURE C.22 PHEV-40: Sensitivity of buydown cost to changes in input variables.

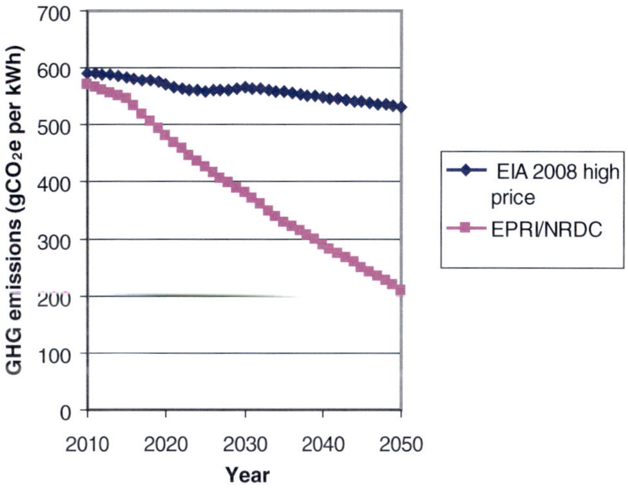

FIGURE C.23 GHG emissions from the future electric grid.

and, especially, the PHEV-40. The PHEV-40 would reach breakeven in 2024 at a total buydown cost of about $25 billion instead of $400 billion. In the high case, the committee used both the probable cost values and the probable market penetration rate. This delays the break-even year for both PHEVs but can result in a lower buydown cost (because of the delay in buying PHEVs until costs have dropped).

With high oil prices or rapid success in meeting DOE's battery goals, break-even years for PHEV-40s could occur 10 to 15 years sooner and the buydown costs would be much lower than in the base case.

LOW-CARBON GRID

The Electric Power Research Institute (EPRI)/Natural Resources Defense Council (NRDC) scenario used to estimate GHG emissions for a future low-carbon grid assumes wide adoption of advanced low-carbon technologies. The cost for charging electricity is assumed to be 8 cents/kWh for nighttime electricity.

Figure C.23 compares the GHG emissions from two future electric grids: the low-carbon EPRI/NRDC case and the EIA business-as-usual Annual Energy Outlook high-price case. For the latter case, GHG emissions were extrapolated beyond 2030, assuming that electricity demand and GHG emissions for electric generation continue to grow at the same rate as between 2006 and 2030.

Figure C.24 shows the hydrogen GHG emissions per unit of fuel energy assumed for hydrogen in the 2008 Hydrogen Report.

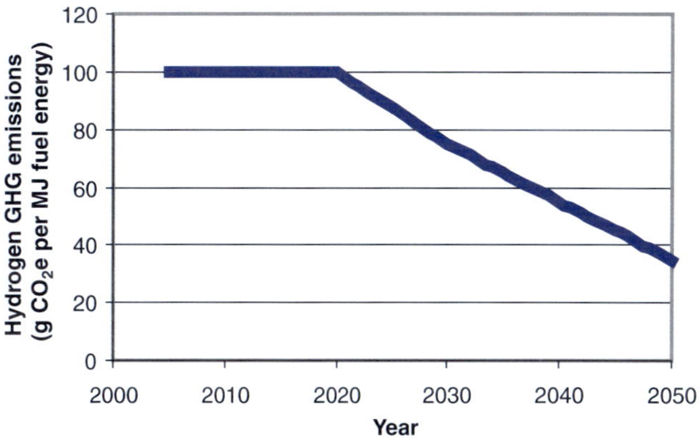

FIGURE C.24 Hydrogen GHG emissions per megajoule of energy.

Appendix D

Statement of Task

Plug-in hybrid electric vehicles (PHEV) are a transportation alternative that was not considered by the Committee on Resource Needs for Fuel Cell and Hydrogen Technologies in its recently issued report *Transitions to Alternative Transportation Technologies—A Focus on Hydrogen.* It is difficult to predict how rapidly PHEVs could penetrate the market because costs, consumer acceptance, and other factors associated with the technology are still very uncertain. Under this contract modification, the committee shall extend its analysis to include the potential impact of PHEVs on petroleum use and CO_2 emissions. The committee shall assess the status of PHEV technology, develop a best-case plausible scenario for how PHEVs may enter the light-duty vehicle market, and use the models developed in its study to estimate the potential reduction of petroleum consumption and CO_2 emissions that might result. Specifically, the committee shall:

(1) Review the current and projected technology status of PHEVs;
(2) Consider the factors that will affect how rapidly PHEVs could enter the marketplace (including the interface with the electric T&D system);
(3) Determine a maximum practicable penetration rate for PHEVs, consistent with the time frame and factors considered in the previous report;
(4) Based on tasks 1, 2, and 3, modify its model appropriately to incorporate PHEVs and develop and estimate impacts on petroleum consumption and CO_2 emissions;
(5) Write a report documenting its study and assessment.

Appendix E

Acronyms and Abbreviations

AEF	America's Energy Future (study)
AEO	Annual Energy Outlook (annual report of EIA)
AER	all-electric range, the distance a plug-in hybrid can travel on battery power alone
BAU	business as usual
BEV	battery electric vehicle
CAFE	Corporate Average Fuel Economy (regulations)
CD	charge-depleting mode of plug-in hybrid operation
CS	charge-sustaining mode of plug-in hybrid operation
DOE	Department of Energy
eff	efficiency
EIA	Energy Information Administration (DOE)
EPRI	Electric Power Research Institute
FCV	fuel cell vehicle
FTD	Federal Test Procedure
gge	gallons gasoline equivalent, used as a measure of energy content
GHG	greenhouse gas
HEV	hybrid electric vehicle
HFCV	hydrogen fuel cell vehicle
HWFET	Highway Fuel Economy Test Driving Schedule
ICEV	internal combustion engine vehicle
kg	kilogram
km	kilometer
kWh	kilowatt-hour (1,000 watt-hours, measure of energy)
MIT	Massachusetts Institute of Technology
NRC	National Research Council
NRDC	Natural Resources Defense Council
OEM	original equipment manufacturer (automobile manufacturer)
PHEV	plug-in hybrid electric vehicle
PHEV-10	plug-in hybrid vehicle with a 10-mile all-electric range
PHEV-40	plug-in hybrid vehicle with a 40-mile all-electric range
Wh	watt-hour (measure of energy)

Appendix F

Estimation of Lithium-Ion Battery Pack Costs

BACKGROUND

Battery pack costs will be the prime determinant of when PHEVs will become competitive, but projections of these costs are highly uncertain. This appendix provides additional detail on the committee's methodology for developing the cost estimates presented in Chapter 2 of the report.[1] The committee reviewed the literature and heard presentations given at NRC meetings by experts. It then discussed its preliminary conclusions about current and anticipated battery pack costs with industry and government experts. Its estimates were subject to the National Academies' report review process during which additional experts commented on the committee's assumptions and analysis. The committee's estimates in its final report include a broad range of costs, reflecting the information the committee reviewed.

Several factors must be known before various battery pack cost estimates can be compared. They include whether the estimated cost per kWh applies to nameplate or usable energy, and, if the former, the assumed state of charge (SOC) range to know how much energy is available, and whether the cost is based on the beginning or end of life capacity (Li-ion batteries deteriorate at about 2 percent per year). To calculate the actual battery pack costs for a vehicle with a specified electric driving distance, the Wh/mile (propulsion energy) required to propel the vehicle must also be estimated, and the cooling (e.g., liquid or air) that is required must be specified. Due to the early state of PHEV development and lack of road experience, there is considerable variability in these cost estimates.

The costs reported in Chapter 2 of the report were based on usable energy, that is, the fraction of the total energy stored in the battery that can be withdrawn for propulsion without risking damage to the battery or causing safety issues. That is the most useful measure when determining the *performance* of the vehicle and is the measure used by the Department of Energy and the U.S. Advanced Battery Consortium in their published goals for battery storage. However, when comparing *costs* of different batteries, assuming equal propulsion energy (Wh/mile), nameplate (or nominal) capacity better describes the battery pack costs actually put into the PHEV. The difference between these two measures for a specific battery pack (essentially the SOC selected by the vehicle manufacturer) has caused considerable confusion. The discussion below of various sources of estimates compares the nameplate cost unless otherwise specified.

Battery lifetime and safety are also very significant PHEV development issues for the industry. The current life expectancy of Li-ion batteries for computers and power tools is around 3 to 4 years,[2] but 10 years or more will be needed for PHEVs to be competitive. R&D programs are reducing battery costs and improving battery durability and safety, but these goals must be met simultaneously and may interfere with each other. The committee concluded that durability and safety goals are more likely to be met than cost goals. Therefore the focus of this discussion is on costs *after* durability and safety problems have been solved.

CURRENT COSTS

The committee reviewed a variety of sources to establish the most probable and optimistic costs for the current generation of battery packs. The review, discussed below, indicated a range of \$500 to \$1500/kWh nameplate. The range is so wide in part because different technologies at different stages of development are reported. Based on this range, the committee selected \$875/kWh as the most probable value and \$625/kWh as an optimistic value for batteries that have already been ordered to be used in the first generation of

[1]This appendix was added to the report after the release of the prepublication version to clarify the basis for the committee's cost estimates. The estimates themselves are unchanged.

[2]The Hymotion kit made by A123 Systems to convert the Toyota Prius to a PHEV comes with a 3-year warranty against defects. The all-electric range is not specifically guaranteed. See http://www.a123systems.com/hymotion/pop_ups/warranty.

PHEV-40s, and $825/kWh and $625/kWh for PHEV-10s. Literature results are as follows:

- The NAS-NAE-NRC report *America's Energy Future* concludes that automotive-grade Li-ion battery pack costs today are between $500/kWh and $1000/kWh nameplate (NAS-NAE-NRC, 2009).
- DOE estimates of current costs are $1,000⁺/kWh usable energy (Howell, 2009).[3] DOE goals are for performance at the end of life. Li-ion batteries deteriorate over time, typically at about 2 percent per year. Assuming a DOE start of life SOC of 70 percent, the committee estimates DOE's nameplate cost at start of life to be $560⁺/kWh.
- A recent McKinsey report concludes that battery pack costs range from $700/kWh to $1,500/kWh nameplate (Hensley et al., 2009).
- A 2009 paper (Shiau et al.) from researchers at Carnegie Mellon University uses $1000/kWh nameplate.
- Pesaran et al. (2007) estimated the cost of advanced Li-Ion battery costs as ranging from $800/kWh to $1,000/kWh nameplate or higher.
- The Zero Emissions Vehicle Report projected a "current" cost of about $500/kWh nameplate in 2006 (Kalhammer et al., 2007).

The following two reports were released after the committee completed its analysis, but they are included here for completeness.

- The Electrification Coalition's 2009 report *Electrification Roadmap* (available at www.electrificationcoalition.org) states that the current cost is $600/kWh nameplate
- The Boston Consulting Group's 2009 report *Batteries for Electric Cars* (available at www.bcg.com/documents/file36615.pdf) says it is $1000-1200/kWh

The committee expects that these early PHEVs will employ a conservatively low SOC, about 50 percent, to ensure battery durability and safety. With experience and improved battery and control technology the SOC may be increased to 70 or even 80 percent, but that is speculation until several years of real-life operating experience indicate whether battery durability would be jeopardized.

At 50 percent SOC, the current cost for usable (or available) energy for a PHEV-40 comes to $1750/kWh (probable) as shown in Chapter 2, and the nameplate cost is $875/kWh. Based on the report's assumed propulsion energy of 200 Wh/mile, a 16 kWh battery pack (8 kWh usable) such as will be used in the Chevrolet Volt costs $14,000. While neither GM nor LG Chem, the battery supplier, has announced the costs, $7000 for the cells has been reported in the media. The additional systems, materials, and labor to assemble a battery pack are substantial.

The committee concluded, based on research and discussions, that the cost of assembling the pack is approximately the same as the cost of the cells, corresponding to the total of $14,000 for the PHEV-40. The committee also estimated a range of costs, recognizing the uncertainty involved, and concluded that under more optimistic assumptions the cost could be $10,000. In comparison, DOE estimates that a PHEV-40 would require 11.6 kWh usable energy in a pack that would cost over $11,600, consistent with the estimating accuracy of this report.

PROJECTED FUTURE COSTS

The committee estimated future costs of Li-ion batteries based on the technology status and cost projections in the literature. Based on this analysis, the committee judged that battery pack costs are likely to decline by about 35 percent by 2020 and 45 percent by 2030, as shown in Tables 2.2 and 2.3. This yields a nameplate 2030 PHEV 40 battery pack cost of about $500/kWh ($1000/kWh usable) or, under more optimistic assumptions, about $360/kWh. The committee did not attempt to estimate the future costs if a major technology breakthrough occurs, such as the development of a durable, safe Li- air battery.

The literature contains a wide range of projected future Li-ion battery and battery pack costs (all costs are nameplate unless otherwise noted):

- The DOE goal is for a very rapid cost reduction from the estimated $1,000⁺/kWh current cost to $500/kWh in 2012 to $300/kWh (all costs based on available energy base) in 2014.[4] Assuming 70 percent SOC and 20 percent deterioration conversion factors, DOE's goals correspond to $280/kWh in 2012 and $168/kWh in 2014 on a nameplate capacity basis. Meeting these goals would result in a $1700 cost for a 3.4 kWh battery pack in a PHEV-10, and $3,400 for an 11.6 kWh pack in a PHEV-40. Note that these are goals, not projections. Meeting these goals could result in PHEVs being competitive in the marketplace much more rapidly compared with HEVs and conventional vehicles, as discussed in Appendix C.
- The U.S. Advanced Battery Consortium (2009) has the same goals as DOE.
- Nelson et al. (2009) projected pack manufacturing cost of about $350/kWh at 100,000 unit volume for a PHEV-10 and $200/kWh for a PHEV-40.
- The McKinsey report projected that costs will decrease at 6 percent to 8 percent per year, and, with aggressive cost reduction, could reach $420/kWh nameplate by 2015 (Hensley et al., 2009).

[3]See also S. Satyapal and P. Davis, presentation to the Committee on Review of FreedonCAR & Fuel Partnership, Phase 3, Washington, D.C., 2009.

[4]T.Q. Duong, Update on electrochemical energy storage R&D, presentation to the committee, Washington D.C., June 2009.

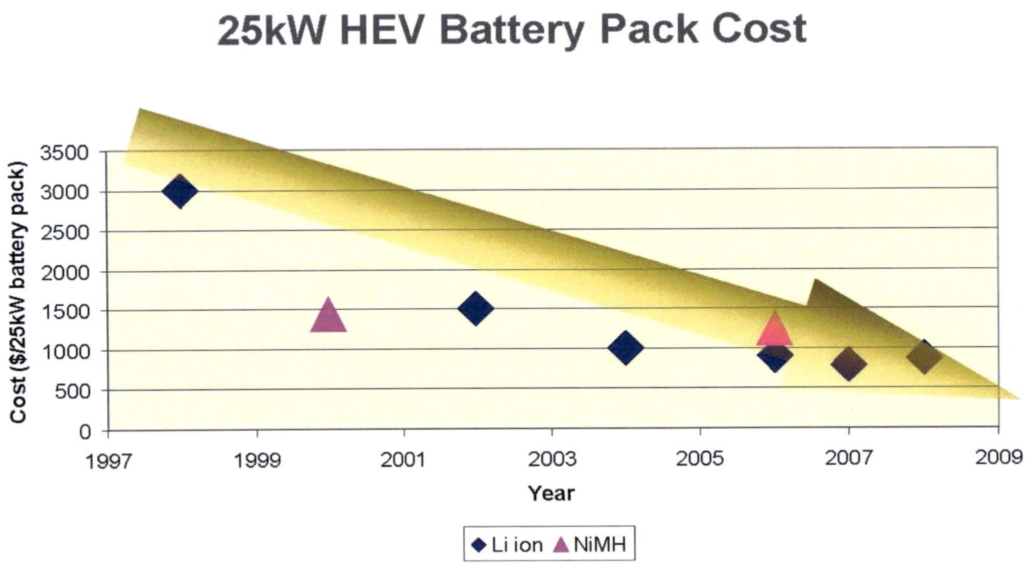

FIGURE F.1 Historical cost reduction experience for NiMH battery packs and for Li-ion battery packs. Recent experience does not suggest rapid further cost reductions. SOURCE: T.Q. Duong, Update on electrochemical energy storage R&D, presentation to the committee, Washington D.C., June 2009.

- Anderman predicts that the cost of Li-ion batteries will remain at around $600/kWh even with increased production (Anderman, 2007).
- Kalhammer et al. (2007) project costs from $350/kWh to $400/kWh (nameplate) for PHEV-40 battery packs at volume production (100,000 to 200,000 units per year). Costs for PHEV-10 battery packs are projected to be $560 to 860/kWh for production at 100,000 to 625,000 units per year.

The future cost estimates in this report are higher than most, but not all, other projections, especially the DOE goals. The committee concluded that reductions greater than 50 percent in battery costs are unlikely over the next two decades without a major technology breakthrough, because meeting battery durability and safety goals could slow cost reductions. For example, raising the SOC range would be a significant cost saver but could compromise durability if that put too much stress on the cells.

Cost reductions are likely to come mainly from improvements in technology, with lesser contributions from manufacturing improvements, improved yield, and manufacturing scale.[5] Technology will continue to improve, but it is already well developed for current Li-ion cells. Cells for automotive applications will be bigger than current Li-ion cells but are otherwise not very different in either chemistry or manufacturing processes. Thus the potential for large cost reductions from technology improvements is limited. Furthermore, materials represent more than half the cell cost (Nelson et al., 2009), and these costs are unlikely to drop dramatically.

Economies of scale are often cited as a factor that can drive down costs, but hundreds of millions to billions of Li-ion cells already are being produced in optimized factories. Building more factories is unlikely to have a great impact on costs. The cost of the battery pack enclosure that holds the cells, the electronics required to monitor and control each cell to prevent over-charging and run-away, and the temperature control system to manage battery pack temperature are a major portion of the total battery pack cost. These components are unlikely to undergo larger cost reductions than the cells, and so the committee maintained the same ratio of twice the cell cost for the pack.

Li-ion batteries have undergone large cost reductions over the last 10 years, but the costs seem to be leveling out. Costs of NiMH battery pack for HEVs have declined only modestly in recent years, as shown in Figure F.1, suggesting that further major cost reductions are not very likely without technology breakthroughs, which this study did not try to project.

The committee considered all these projections and other information to come up with its estimates for 2030 future costs of about $500/kWh or, under more optimistic assumptions, of about $360/kWh.

[5]According to one estimate, cell costs could drop by more than 50 percent by 2015, with almost all of that decrease coming from technology and process improvements (D. Vieau, A123 Systems, presentation to the committee, Washington, D.C., May 2009).

VALIDATION OF COSTS

Based on the above and other related information, discussions with industry and government experts, and its own judgments, the committee developed Tables 2.2 through 2.6 with what it considered the most probable set of numbers for the key battery pack performance parameters and battery pack cost. These were discussed with the companies acknowledged in the front matter of this report to ensure that battery costs and performance were realistic.[6] Some adjustments were made based on these discussions, and the estimates presented above were finalized. The committee's report was subjected to the National Academies' report review process in which another set of experts critiqued the report and the committee's assumptions and analysis.

[6]Toyota Motor Corporation, General Motors, Ford Motor Corporation, A123 Systems, Compact Power Inc. (LG Chem), Delphi Corporation, DENSO International America, Inc., U.S. Department of Energy.